TOWER
BRIDGE

1894 to date

COVER IMAGE: **Tower Bridge.**
(Ian Moores – ianmooresgraphics.com)

First published in May 2019

A catalogue record for this book is available from the British Library.

ISBN 978 1 78521 649 7

Library of Congress control no. 2019934273

Published by Haynes Publishing,
Sparkford, Yeovil,
Somerset BA22 7JJ, UK.
Tel: 01963 440635
Int. tel: +44 1963 440635
Website: www.haynes.com

Haynes North America Inc.,
859 Lawrence Drive, Newbury Park,
California 91320, USA.

Printed in Malaysia.

Senior Commissioning Editor: Jonathan Falconer
Copy editor: Michelle Tilling
Proof reader: Penny Housden
Indexer: Peter Nicholson
Page design: James Robertson

TOWER BRIDGE

1894 to date

Operations Manual

Insights into the history, design, construction and operation of this London icon

John M. Smith

Contents

BELOW Tower Bridge's high-level walkways. The outer catenary cable of the west high-level tie suspension bridge can be seen. *(Author)*

ABOVE The view downstream towards Canary Wharf from the east high-level walkway. *(Author)*

Introduction

When I was a boy I spent many happy hours poring over my Dad's boyhood copy of *The Wonder Book of Engineering Wonders*. Tower Bridge was the first presented in the chapter on famous bridges. Some 30 years later I visited the bridge for the first time and realised it was actually a steel bridge, clad with a thin layer of masonry. I marvelled at the enormous fabricated steel beams, each with row upon row of beautifully executed rivets. The magnificent steam pumping engines also impressed. I vowed that, one day, I would write a book about the bridge.

The editor of *The Wonder Book of Engineering Wonders* was right to include the bridge as it was, and it remains an engineering marvel. One of the first steel bridges in the world, it employed cutting-edge steam and hydraulic technology. It is actually ten bridges in one – two masonry arch bridges forming the approaches, two suspension bridges linking the abutments with the river piers, two simply supported steel bridges carrying the roadway over the bascule chambers, two bascule bridges and two cantilevered footbridges. Since 1960, when catenary cable suspension bridges were added to carry the weight of the high-level ties of the suspension bridges, there have been 12 bridges on the site.

During the first half of the 19th century, the

BELOW London's docklands burning during the Blitz in 1940. *(Getty Images 3315052)*

population of London more than doubled, fuelled by the success of the city as the world's busiest port. This put enormous strain on London Bridge, which became a notorious pinch-point for traffic heading to and from the docks. The need for a new bridge near the Tower of London was foreseen as early as 1824, when a high-level suspension bridge was proposed by Captain Samuel Brown RN and civil engineer James Walker, but it would be another 60 years before the Court of Common Council, the primary decision-making body of the Corporation of London, authorised the building of the bridge.

The initial architectural design for the bridge was by Horace Jones, Architect and Surveyor to the City of London and Clerk of the City's Works. The designer and engineer of the bridge was John Wolfe Barry, supported by Henry Marc Brunel and George Stevenson. Builders of the bridge included Sir William Arrol, who also built the Forth Bridge, and Baron Armstrong of Cragside, a veritable titan of 19th-century engineering.

After an eight-year construction period, Tower Bridge opened to road and river traffic on 30 June 1894. It immediately became as much an icon of London as St Paul's Cathedral and the Tower of London, despite being built some 200 and 800 years later respectively. Today, all three are Grade I listed buildings.

The bridge survived Zeppelin raids in 1915–16 and, during the Second World War, the image of the bridge outlined against a burning city in the Blitz of 1940 is particularly poignant. Contrast this with the photograph that shows the bridge on a lovely summer's day during the heatwave of 2018.

The bridge has been struck several times by shipping, the 7,723-ton MV *Monte Urquiola* striking the bridge at least four times, with one Thames waterway official saying 'we're getting

BELOW Tower Bridge on a hot summer's day in 2018. *(Author)*

sort of used to it'. It wasn't personal. *Monte Urquiola* also struck John Rennie's London Bridge on 27 January 1954. She also collided with a Staten Island ferry in 1960 and, 9 miles off Beachy Head, an American freighter in 1961.

The bridge has escaped being hit by aircraft on at least seven occasions, the most dangerous of which was on 5 April 1968 when Flt Lt Alan Pollock flew his Hawker Hunter jet fighter aircraft through the bridge in an unauthorised celebration of the 50th anniversary of the RAF. Arrested upon landing, he would later be invalided out of the RAF to avoid the publicity of a court martial. Thankfully, the only other ill effect of this incident was that a cyclist fell off his bicycle in surprise and tore his trousers.

Another notable incident occurred in December 1952 when the bridge opened while a No. 78 double-decker bus was crossing from the South Bank. The quick-witted driver, Albert Gunter, put his foot down, clearing a 1m gap in the road and dropping nearly 2m on to the opening Middlesex leaf. The bus chassis was bent and there were a few injuries, including a broken leg suffered by the conductor. The City gave Albert a day off work and the princely bonus of £10 for his pains.

During its lifetime, the bridge has been raised and lowered nearly half a million times, always giving precedence to shipping rather than road traffic, as witnessed by the incident in May 1997 when the motorcade of US President Bill Clinton was divided by the opening of the bridge.

Tower Bridge is now much more than a bridge. It is a major tourist attraction, a museum, a popular film location, a concert venue and a prestige location for wedding receptions and banquets. A highlight of the bridge for visitors, the glass walkways give a frightening insight into what it must have been like for the more than 400 men who built the bridge in the days before health and safety became a major consideration on construction sites.

This book covers the history of the bridge from its inception to the present day, covering the selection of the eventual design, the construction phase and the men who built the bridge. It also offers an engineering appreciation of the bridge. As it was constructed in Victorian times, Imperial units are generally used in the description of the bridge and in the very few calculations presented. Appendix 3 includes conversion tables for readers more accustomed to SI units.

I thank Colin Buttery of the City of London for permission to licence images owned by the Tower Bridge organisation and for allowing me to publish my personal photographs of many parts of the bridge. I am also very grateful to Dr Graham W. Owens for his many helpful comments on the manuscript and for reviewing the calculations.

I hope that this Haynes Operations Manual will enable you to fully appreciate the engineering marvel that is Tower Bridge.

John M. Smith
BSc (Hons) MSc CITP CEng FBCS FIET,
Maidenhead, 2019

BELOW A glimpse of the bridge is always a thrill for visitors flying into or from London's Heathrow airport.
(Author)

OPPOSITE MV *Monte Urquiola* getting up-close and personal with London Bridge.
(Press Association)

Chapter One

Build us a bridge!

───●───

From 1800 to 1860 London's population tripled. Pedestrians and goods travelling to and from the docks caused gridlock on London Bridge. Pressure mounted for a new Thames crossing east of the Tower of London – a 'Tower Bridge'. The Special Bridge or Subway Committee was established to evaluate options. Enter many hopefuls and two great men – Sir Joseph Bazalgette and Horace Jones.

OPPOSITE Toil, glitter, grime and wealth on a flowing tide, by **W.L. Wyllie, 1883.** *(Tate Images)*

Victorian London – confident and thriving

In the 1850s, Great Britain was the industrial and economic powerhouse of the world. Her empire was expanding, with territorial acquisitions in Africa a primary objective. The Royal Navy dominated the seas, protecting Britain's trade with all parts of the globe. Gunboat diplomacy was popular, because it worked. The British were not averse to the odd war to protect our interests and send a powerful message. After all, the wars were short and Britain always won.

During the first half of the century, Britain's exports of all manner of manufactured goods had seen a fivefold increase, with textiles, coal and iron accounting for over 50% of commodity exports. London was the world's largest port and wealthiest city. It was the hub of international finance, a centre of shipbuilding, manufacturing, trade and culture, and the beating heart of the British Empire.

We do not have to look beyond the works of Dickens and Conan Doyle to know that Victorian London had a dark side. Poverty and squalor sat side by side with extreme wealth and privilege, as witnessed in Charles Booth's 1889 'Descriptive Map of London Poverty'. London could also be a dangerous and unpleasant place. While the crime rate was generally falling due to the efforts of the London Metropolitan Police Force (created in 1839), white-collar crime and company fraud were rife. Widespread fear was caused by a spate of muggings, during which victims were approached from behind and garroted, prior to being relieved of their valuables. Jack the Ripper was waiting in the wings.

There were other dangers. In his 1842 report on the sanitary condition of the labouring population, Edwin Chadwick – later Sir Edwin – reported that for the year ending 31 December 1838 the average age at death for Whitechapel residents was 45 years for professionals and gentry and their families, while for tradesmen and their families it was 27 years. For labourers and servants and their families it was just 22 years. These numbers were skewed by the horrifically high level of child mortality; around 55% of all children born died before their fifth birthday. The biggest killers were typhus, scarlet fever, smallpox, diphtheria, cholera and consumption (tuberculosis).

Methane venting from London's 200,000 cesspits mingled with smoke from coal fires to form an atmosphere far deadlier and more fog-inducing than the particulate and nitrogen oxide-laden atmosphere of the city today. Cesspits leached into the water table with disastrous public health consequences, including cholera outbreaks in 1831/32, 1848/49 and 1853/54, in which some 31,500 people died.

By 1850, 'railway mania' was over, but 6,000 miles of permanent way had been built in Britain, linking London with the industrial heartlands of the West Midlands, Manchester

and Liverpool, Sheffield, Leeds and Newcastle, and providing goods transportation volumes and speeds vastly superior to those offered by the canal network. New railway termini encircled the city – Euston (1837), Paddington (1838), Fenchurch Street (1841), Waterloo (1848), King's Cross (1850) and St Pancras (1863). Change happened quickly; major projects were conceived and implemented in just a few years – not decades.

The Great Exhibition of 1851 in Hyde Park provided a showcase to the world of Britain's prowess in science and technology, invention, engineering, manufacturing and culture. The Crystal Palace, the central icon of the exhibition, reflected the confidence and dynamism of the period and made a powerful statement that Britain – and London – were open for business.

Toil, glitter, grime and wealth on a flowing tide

The Pool of London extended east from London Bridge to Cherry Garden Pier (the Upper Pool) and from Cherry Garden Pier to Limekiln Creek (the Lower Pool). For centuries, goods were traditionally ferried by fleets of lighters between ships moored in the Pool and the wharves lining both banks of the river.

By 1800, congestion on the river had become so extreme that it limited the ability of the port to expand. This led to a massive off-river dock-building programme, during which the West India Docks (1802), East India Docks (1806), the Surrey Commercial Docks (1807), the London Docks (1815), St Katharine Docks (1828), Victoria Dock (1855), Millwall Dock

BELOW The Great Exhibition: moving machinery, by Louis Hague, 1806–85. *(The Royal Collection)*

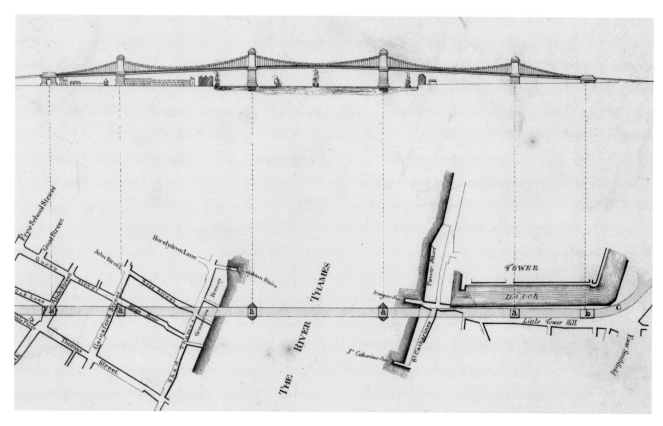

ABOVE The 'bridge of suspension' proposed by Captain Samuel Brown RN and James Walker, civil engineer. *(Science Museum)*

(1868) and the Royal Albert Dock (1880) were constructed. These provided the ability for ships to unload directly into purpose-built warehouses, giving the port a quantum leap in capacity but depriving lightermen and wharfingers in the Pool of London of their livelihoods.

London's traffic problems

In 1801 the population of London had been 1.1 million. In 1860 it was 3.2 million, one-third of whom lived to the east of London Bridge. Streets near the bridge approaches were gridlocked with horse-drawn vehicles of all descriptions, including omnibuses. The problem was largely caused by the success of the new docks. Goods could now be landed easily, but transporting them by road to the railway termini was a huge challenge.

OPPOSITE The entrance to the Thames Tunnel linking Rotherhithe with Wapping, by B. Dixie. *(London Metropolitan Archive – LMA/ Collage 22412)*

The need for a new bridge, downstream from London Bridge, was foreseen as early as 1824 by Captain Samuel Brown RN. With civil engineer James Walker, Brown proposed a high-level suspension bridge running from the Minories to Bermondsey. The location proposed for the bridge was that occupied by Tower Bridge today. The bridge is shown in the image above. It is believed to be the earliest proposal for a bridge just east of the Tower of London.

The Thames Tunnel, designed and built by Marc Brunel and his son Isambard Kingdom Brunel to link Rotherhithe and Wapping, opened to pedestrian traffic in 1843. It provided a means to cross the river for those able to afford the one penny toll and fit enough to walk down 80ft of stairs, through the tunnel and up another 80ft of stairs on the other side. American visitor William Drew described his experience of using the tunnel: 'Ladies in fashionable dresses and with smiling faces wait within and allow no gentleman to pass without giving him the opportunity to procure some pretty thing. …'

Unfortunately, a good number of these affable ladies were selling services rather than goods, the tunnel earning a very bad reputation. It was a commercial failure and sloping approaches to permit the tunnel to be used by horse-drawn carriages were never constructed. By December 1869, trains were running through the tunnel.

Another tunnel, the Tower Subway, was constructed in 1871, designed to carry passengers through the tunnel on a narrow-

BUILD US A BRIDGE!

gauge railway. The venture foundered and the tunnel was adapted for foot passengers; it carried over 1 million passengers per year at a toll charge of a halfpenny. The subway closed in 1898, following the opening of Tower Bridge, but is still in use today as a communications tunnel.

The Engineer was a weekly magazine for engineers, inventors and technicians. Its editor did not shrink from expressing a view on any topic of the day. The edition of 28 January 1876 expressed a typically robust clarion call regarding London Bridge and City traffic:

We have heard a good deal lately of the block on London Bridge and the insufficiency of that structure for the wants of the day. But so far no one has been content to examine whether the unfortunate bridge is or is not justly made odious for the sins laid to its charge. For ourselves, we think that the approaches to the bridge are every bit as much to blame, and at the same time are more susceptible to improvement. If the approaches to a bridge be cramped the bridge will be crowded be it ever so wide. Take Fenchurch Street for example, the first direct street after the bridge is crossed which has to support the traffic to the docks. At the present time there are, in that street, holes inches deep in the asphalt filled in with cubes of granite. There is only room for one line of vehicles each way, and a momentary stoppage of an omnibus blocks the traffic sometimes all the way to the bridge. Again more horses fall between Philpot-lane and Lombard-street than in any other part of London. If the approaches to the bridge were dealt with by some comprehensive scheme we should hear less of obstruction on the bridge. The approaches are, however, not the only offenders. We allude to public vehicles. Omnibuses start from the Mansion House for instance, and crawl in the most deliberate manner in a procession as far as Ludgate-hill, stopping, as a rule, at all chief corners, and picking up passengers as they go. This should not be. The convenience of the City at large should not be sacrificed to what the omnibus proprietors conceive to be their best interests. If it were insisted that, at all events, a large majority of these

vehicles should not stop except at fixed and somewhat distant intervals, as is the rule in Vienna or Paris, the traffic would be facilitated and the interests of the proprietors would not suffer.

In August 1882, traffic across London Bridge was observed over a two-day period – presumably during the working week. The average daily traffic was 22,242 vehicles and 110,525 pedestrians. We do not have details of the types of vehicle but can assume that the vast majority were horse-drawn, including omnibuses, and that a very few steam-powered road locomotives might have been included in the mix.

The bodies responsible for London's bridges

For centuries, bridges in the City of London had been the remit of the Bridge House Estates Trust of the City of London Corporation, administered by the Bridge House Estates Committee (BHEC). Bridge House Estates was established by Royal Charter in 1282 with responsibility for the maintenance of (old) London Bridge, Bridge House having been the administrative headquarters of old London Bridge, situated near St Olave's Church in Tooley Street, Southwark.

BHEC was originally funded by the rents and leases of the buildings on old London Bridge and by tolls levied on users of the bridge. By 1855, the responsibilities of the committee extended to other bridges in the city and the committee administered a large fund which could be allocated to new projects. Bridge House Estates funded the construction of the new Blackfriars Bridge, designed by the great railway engineer Joseph Cubitt, which opened in November 1869 at a cost of £350,000.

In 1848, the Metropolitan Commission of Sewers (MCS) was established. The Building Act of 1844 required that all new buildings be connected to a sewer, and the commission set about the task of connecting existing cesspools to sewers, which flowed directly into the river. This was a disastrous policy.

The Metropolitan Board of Works (MBW) was

created by the Metropolis Local Management Act of 1855 to regulate the construction of buildings in London and to provide the improvements in infrastructure required to cope with London's rapid growth. It was an appointed, rather than elected, body; this did not sit well with rate-payers. There was a full-time chairman and 45 members, of which 3 were proposed by the City of London and 42 were appointed by parish vestries. The powers, duties and liabilities of the board would later pass, by the 1888 Local Government Act, to the London County Council.

The disposal of human waste was a primary focus of the MBW, it having taken on the responsibilities of the MCS upon establishment. Matters came to a head in 1858, the year of 'the great stink'. The board's chief engineer, Joseph William Bazalgette, launched a massive project to design and build a core sewer system by which sewage from 1,000 miles of new main and subsidiary sewers would be transported by gravity in six main interceptor sewers, 82 miles in total length, to the Thames Estuary, where it would be pumped to the surface, held in covered reservoirs and released from northern and southern outfalls into the river on the

ABOVE **Old London Bridge from the west, by Claude de Jongh, 1650.** *(Victoria & Albert Museum – V&A – Image 2006BG2028)*

LEFT **Joseph William Bazalgette, chief engineer to the Metropolitan Board of Works.** *(LMA/ Collage 14155)*

LONDON BRIDGE. 7171 Ⓡ. J.V.

high tide.[1] Bazalgette demonstrated excellent judgement in sizing the interceptor sewers; he calculated the diameter needed to cope with the worst case – and then doubled it, giving the system four times the worst case capacity.

Subsequent Acts extended the responsibilities and powers of the board to parks and open spaces, and to the improvement of streets and bridges.

So now there were two bodies with responsibility for bridges – and they did not always see eye to eye. For example, a Bill was introduced in June 1864 that would authorise MBW to purchase Southwark Bridge and other toll bridges and open them to the public free of charge – with some of the money to come from Bridge House Estates funds. BHEC saw this as both unprecedented interference in their business and an unjust raid on their funds. They successfully opposed the Bill.

---

1 In truth, this system just pushed the problem further downstream. In 1900, treatment works were built at the outfalls to substantially reduce pollution in the Thames Estuary.

Demand grows for a new river crossing east of London Bridge

On 12 June 1868, the Corporation of London *did* purchase Southwark Bridge, after a trial period during which it had been operated as a toll-free bridge by the bridge company in return for financial contributions from the Corporation. While vehicular traffic carried by Southwark Bridge had initially leapt from 1,000 to 3,000 vehicles per day when the tolls were removed, it was failing to significantly relieve the pressure on London Bridge. The principal reason for this was that the steeply inclined approaches to Southwark Bridge deterred horse-drawn traffic.

A report to BHEC in 1871 by Col Haywood, Engineer to the City Commissioners of Sewers, recommended a new crossing east of London Bridge, but no tangible steps were taken to progress the recommendation.

In 1874, the Court of Common Council – the primary decision-making body of the Corporation of London – instructed BHEC to make recommendations on the best way to increase the capacity of London Bridge. In December 1875, the committee reported, recommending that London Bridge be widened, but pressure was growing from business houses, merchants and traders for an entirely new crossing east of the Tower of London – a 'Tower Bridge'.

In January 1876, the court instructed BHEC to consider the relative advantages and costs of a new bridge or subway at this location. Due to the high workload of BHEC, a new committee was formed in February 1876 – the Special Bridge or Subway Committee (SBSC) – to consider and weigh up the merits of alternative approaches to the crossing. At the inaugural meeting of the SBSC on 17 February 1876, Henry Aaron Isaacs was elected as chairman. He would go on to become Lord Mayor of London in 1889. It was resolved that the committee would meet on the second Monday of each month (with the exception of August) at 1.00pm. Momentum was building.

Possible solutions

J.E. Tuit[2] neatly summarises the possible approaches to the provision of a new river crossing east of the Tower of London:

- A low-level bridge with uninterrupted spans.
- A low-level bridge which can be opened to allow the passage of vessels.
- A high-level bridge with inclined road approaches, providing sufficient clearance for high-masted vessels to pass beneath it, even at high tide.
- A high-level bridge with hydraulic lifts at each end.
- A tunnel under the river, with inclined approaches.
- A tunnel with hydraulic lifts at each end.
- A ferry.

2   J.E. Tuit, *The Tower Bridge – Its History and Construction*, Smith Elder & Co., 1894.

A low-level bridge would be ideal for vehicular traffic and pedestrians, but would be strongly opposed by the wharfingers conducting business in the reach between the proposed bridge and London Bridge; the new bridge preventing all but the smallest vessels from reaching their wharves. Such opposition would have to be overcome by means of compensation for lost livelihoods.

A low-level bridge provided with a means of opening to allow the passage of vessels would be satisfactory provided there was sufficient width in the opening section to permit large vessels to pass safely. Difficulties were foreseen, however, as congestion in the Upper Pool was such that there were three or four tiers of ships on each side of the river, leaving a very restricted clear channel.

High-level bridges or tunnels with long inclined approaches were not favoured as additional land would need to be purchased for the approaches and heavy horse-drawn vehicles would have difficulty in negotiating the slopes safely.

Bridges or tunnels with hydraulic lifts at both ends would remove the need for additional land for approaches, but would add precious time to the crossing. The need for the lifts to run non-stop, would imply substantial operating expenses too.

Tunnels were not in favour. The tunnel under the Mersey had taken 6 years to complete and that under the Severn 13 years. Closer to home, Brunel's Thames Tunnel had been beset with problems, the workings being inundated six times and the tunnel taking 18 years to complete (including 6 years for which the workings were closed owing to funding problems). Construction difficulties could be expected, but not accurately foreseen, and the capital and ongoing maintenance costs of pumping equipment to handle the inevitable seepage was high.

Ferries were not viewed as a satisfactory solution to the need for a 24/7 crossing. It was quite normal for traffic on the river to be suspended for 30 days or so per year owing to fog. Added to this were some 12 days of severe frost which prevented shipping movements, so ferries were not seen as the answer.

From the ridiculous to the sublime

Proposals considered by the Special Bridge or Subway Committee, June 1876–December 1878

The SBSC was not well equipped to evaluate competing bridge designs and make a recommendation. The committee included elected members and professional men, but they were not engineers. Neither were they experienced in contracting or project management.

Although they did ask the Court of Common Council for permission to advertise for bridge designs, this permission was not forthcoming. However, it seems that news of the committee's work spread around the civil engineering community and letters from budding bridge designers to the committee, or to the Lord Mayor, would result in an invitation for the sender to attend a meeting of the committee to present their design.

So it was that, at the meeting on 13 June 1876, the Thames Steam Ferry presented its plans for a regular service. Others followed:

1 On 18 July 1876, civil engineer John Keith presented his proposal for a cast-iron arch subway or 'sub-riverian arcade' (below). This would be built upon a concrete foundation below the river bed. His cost estimate was £343,600 to build the subway and £100,000 for land acquisition. The picture conjures up a lovely vision. What could be more pleasant than spending a little time in the arcade of shops? Of course, with so many horses passing by, the reality would not have been quite so delightful. The top of the cast-iron subway is shown on Keith's drawing (out of view in the image below) to be perilously close to the river bed.

2 On 26 July 1876, George Barclay Bruce presented his idea of a transporter bridge (opposite) to the committee. Bruce was a respected engineer who had served an apprenticeship with Robert Stephenson and worked on several railway projects. Six piers, dividing the river into equal 100ft channels were each to be equipped with independent powerplant driving rollers. A bridge platform 300ft long and 100ft wide and weighing some 5,000 tons when fully loaded with 100 vehicles and 1,400 pedestrians would shuttle to and fro between the banks, always supported by the rollers on at least two and sometimes three piers. The bridge would leave the same shore every six minutes. Bruce's

RIGHT John Keith's 'sub-riverian arcade'. (LMA COL/SVD/ PL/03/0271)

LEFT **George Barclay Bruce's transporter bridge.** *(LMA COL/ AC/16/AS551)*

estimate of the capital cost was in the region of £134,000 plus £10,000 for expenses, which seems remarkably low for such a complex arrangement. It would have been terrifying for those on a vessel passing through one of the channels between the piers to see 5,000 tons of steel, humanity and horseflesh looming towards them out of the fog, and very disconcerting for passengers on the platform should it jam halfway across the river.

3 Also on 26 July, inventor Frederic Barnett pitched to the committee his proposal for a duplex bridge (below). This was a double swing bridge designed to allow shipping and vehicular traffic to pass unimpeded at all times. The idea that the opening of the two swing bridges could be choreographed sufficiently well to allow the passage of sailing ships at the mercy of the tide and the wind, while safely switching horse-drawn traffic and

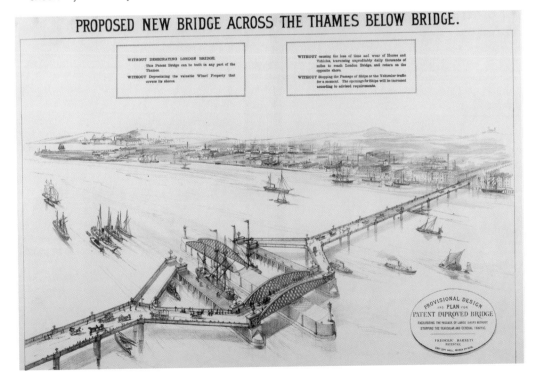

PROPOSED NEW BRIDGE ACROSS THE THAMES BELOW BRIDGE.

LEFT **Frederic Barnett's duplex bridge.** *(LMA/Collage 22768)*

hurrying pedestrians from one carriageway to the other, sounds extremely risky. The cost was estimated at between £400,000 and £500,000.

4   On 11 September 1876, Mr Edward Perrett presented his proposal for a high-level bridge to the committee. The bridge had three 267ft spans and an 80ft headway above Trinity high water. Hydraulic hoists would be installed at each end and stairs would be provided for foot passengers. Each hoist would be capable of making 30 journeys per hour. The estimated build cost was £340,000 and the annual operating cost was expected to be £4,000.

5   At the same meeting, engineer Thomas Claxton Fidler described his proposed bridge to the committee. It was to be a three-span high-level bridge with a headway of 70ft above Trinity high water. Northern access would be by means of a hydraulic lift. The southern approach was to be by means of a spiral. The estimated build cost was £380,000, plus £110,000 for property acquisition and compensation. Fidler would later publish a book on bridge construction.

6   Also at the 11 September meeting, Sidengham Duer pitched his proposal for a high-level bridge with a headway of 80ft above Trinity high water (below). To eliminate the need for long approaches, the bridge would be equipped with a hydraulically operated platform at each end. The estimated cost was £125,000 plus an annual operating cost of £1,900. Duer would later design hydraulic lifting docks.

Letters from John Keith and Sidengham Duer were read out at the meeting on 9 October 1876. Two further proposals had also been received, one by John P. Drake for a swing bridge, and one from C.T. Guthrie for a transporter bridge or 'River Railway Line'. This featured a level track on the bed of the river, on which a self-propelled platform would shuttle between the Middlesex and Surrey banks. The estimated cost was £30,000.

The committee was being swamped. It turned to City Architect, Horace Jones, who was requested to prepare an evaluation of the unsolicited designs submitted so far. This he did.

BELOW Sidengham Duer's high-level bridge. (LMA COL/SVD/PL/03/281)

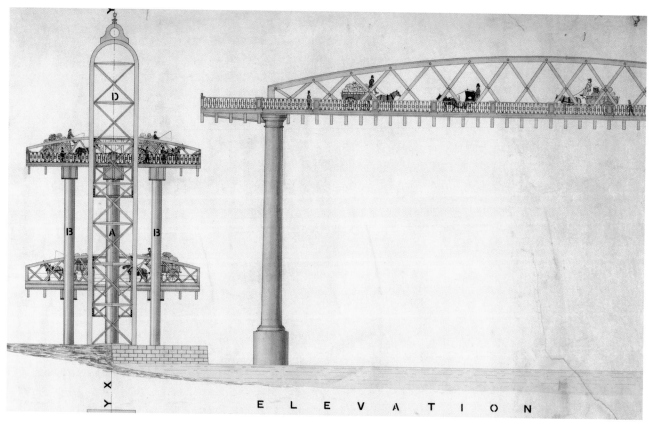

On 13 November 1876, it was resolved that the City Architect be requested for a more detailed report. It was also resolved that:

… it is the opinion of this Committee, that provided the requisite funds can be obtained for the purpose, it is desirable that a Bridge over or a Subway under the River Thames should be constructed Eastward of London Bridge and that the most eligible site will be that approached from Little Tower Hill and Irongate Stairs on the North side and from Horselydown Lane and Stairs on the South side of the River.

And it is resolved that a Report be presented to the Court of Common Council accordingly and recommending that it should be referred back to this Committee to consider the best means of carrying the same into effect and of procuring the requisite funds for the purpose and that this Committee should be authorized to advertise for designs and to offer Premiums for those most approved.

At the meeting on 1 February 1877, Henry Isaacs was re-elected as chairman of the SBSC. An order from the Court of Common Council was read out. This required the committee to obtain information as to the gradients and length of approaches necessary for either a bridge or a subway.

As was always the case when any heavy lifting was required, the committee resolved that the City Architect be tasked to report on the matter of gradients and approaches, to specify what headways were needed for high-level and low-level bridges, how deep a subway would need to be and what the estimated costs would be for all of these variants. Horace Jones was required to consult with Stephen William Leach, Engineer to the Conservators of the River Thames, and report back fully to the committee. A year had passed and not a lot had been achieved. Was the SBSC really adding value?

The City Architect's report was read to the committee at its meeting on 26 February 1877. Jones was asked to provide an additional appendix covering the gradients of the approaches to all bridges over the Thames in London.

At the meeting of 19 March 1877, Percy Westmacott presented to the committee on behalf of Captain Douglas Galton a proposal for a hydraulic swing bridge. At the same meeting, Mr G.L. Shand also presented his ideas regarding a swing bridge.

At next month's meeting, on 26 March 1877, Mr C. Piehliere tendered his design for a low-level bridge with a central lifting span to the committee. The two sections of central span could be lifted in five minutes and would provide a headway of 75ft above Trinity high water. The cost would be £300,000, excluding the approaches.

At the meeting on 23 April 1877, the City Architect's report on construction costs was read and it was resolved that Jones should amalgamate this report with the one submitted on 26 February, and that a report should be presented to the Court of Common Council, based on Jones's reports. A pitch by Henry Vignoles on his design for a three-span high-level bridge with a headway of 85ft above Trinity high water was also given. The proposal included the erection of a large warehouse on the Surrey bank around which the southern approach road of the bridge would wind. Total estimated cost was £600,000 including the warehouse. The illustration above shows that the proposed site for this bridge was upstream of the Tower.

At the meeting on 3 May 1877, the report to the Court of Common Council was signed. The SBSC concluded that:

Having very fully and carefully considered this very important question, both with regard to the imperative necessity for the relief of the traffic of this City, and the convenience of the

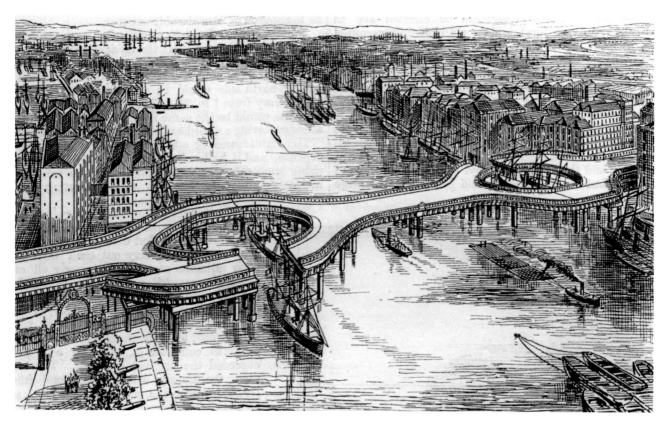

ABOVE Francis
Ingram Palmer's
double duplex bridge.
(Institution of Civil
Engineers – ICE)

*river navigation, we are of the opinion that
the best means to be adopted, in order to
meet the wants of, and to relieve generally
the continuously increasing traffic of this City,
will be to construct at the site indicated in
our former report, viz., that approached from
Little Tower Hill and Irongate Stairs on the
north side, and from Horselydown Lane and
Stairs on the south side of the river, a low-
level bridge, with proper arrangements for
affording the requisite facilities for passage of
vessels up and down the river Thames.*

This report was accepted by the Court of
Common Council. It would seem that some
progress had been made. It would be a bridge
and not a subway!

At the meeting on 9 July 1877, the order
from the Court of Common Council endorsing
the recommendation for a low-level bridge was
read out. George Rennie presented his plan
for a hydraulically powered swing bridge to the
committee and Francis Ingram Palmer pitched
his proposal for a double duplex bridge (above)
with four moving platforms – another proposal
endeavouring to achieve uninterrupted river and
bridge traffic.

It is difficult to understand why the
committee was still providing a platform for
bridge designers and pundits.

In the meantime, the Thames Steam Ferry
had opened for business, plying between
Wapping and Rotherhithe with one boat, the
Jessie May, capable of accommodating 12
two-horse vans. A second boat, *Pearl*, was
on order. The vessels could be propelled in
either direction and were equipped with a
hinged gangway at each end, by which horses,
vehicles and foot passengers were transferred
on to platforms which could be raised to street
level by means of hydraulic rams.

The Metropolitan Board of Works weighs in

Joseph Bazalgette was knighted in 1875
just as he was emerging from a 20-year
period of frenetic activity for the Metropolitan
Board of Works, during which the London
sewer system had been designed and
built, and the Albert, Victoria and Chelsea
Embankments had been constructed. The
Victoria Embankment protected not only one
of the new interceptor sewers but also an

underground railway (the Inner Circle – now the Circle and District Line).

Having finally a little time to devote to other projects, Sir Joseph turned his attention to the matter of the 'Tower Bridge'. What material should be used in its construction, where should it be located and what form should it take?

The material to be used

He started by considering the best material to use for the construction of the bridge. Wrought iron was the traditional material, and British bridge engineers using wrought iron generally adopted a maximum safe stress of 5 tons/in^2 in their designs.

Until very recently, steel had been much more expensive than wrought iron and had been of very variable quality. However, manufacturing improvements made by Dr Siemens and others had both improved quality and reduced cost.

In the design of the St Louis Bridge, constructed of steel, a maximum safe stress of 13 tons/in^2 had been used. This was a dramatic step forward, allowing the ratio of the weight of the supporting structure to the live load carried by it to be dramatically reduced. Steel was Sir Joseph's material of choice.

The preferred location for the bridge

On 10 December 1877, Sir Joseph had reported to the Works and General Purposes Committee of the Metropolitan Board of Works that, in his judgement:

Little Tower Hill, between the Tower and Saint Katharine's Wharf, on the north side of the Thames, rather more than half a mile below London Bridge, is the position best suited for the new bridge, and that the approaches to any other bridge at a lower point would be very inferior to those obtainable near the Tower, and would be seriously interrupted by the St Katharine and London Docks, with their opening dock entrances.

This report included a statement by Mr Leach, the Engineer to the Thames Conservators, objecting strongly to a low-level bridge with a means of transit for shipping through it. In Leach's opinion, there were two serious issues with this proposal. Firstly, in a space crowded with shipping and in a strong wind or tideway, it would be almost impossible to stop large vessels during the opening of the bridge, and then safely to navigate through it. The ships would lose their steerage-way and swing round upon the bridge. Secondly, the interruption of road traffic up to 48 times daily for the passage of vessels would seriously prejudice the usefulness of the new metropolitan thoroughfare.

The preferred type of bridge

It would seem that Sir Joseph agreed with Mr Leach's analysis. He turned his attention to the specification of a high-level bridge. To answer the question of how much headroom for shipping should be provided, Sir Joseph relied upon a study which had been undertaken between 4 and 11 May 1876. During this period, the mast heights of all vessels passing up river westward of St Katharine Docks had been recorded.

Bazalgette concluded that:

if masted vessels which have topmasts exceeding 65 feet above the water level, being one fourth of the total number, would lower their top masts, a high-level bridge might be constructed which should give a clear headway for all ships to pass under it, and which might be provided with convenient approaches and gradients.

A clash of titans

Following the receipt by the MBW of a letter from Mr John A. Brand, the Comptroller of Bridge House Estates Trust, the SBSC met with the Works and General Purposes Committee of the MBW on 26 November 1877. The purpose of the meeting was to determine whether there was scope for the City and the MBW to collaborate on the bridge project. After the meeting, the MBW committee resolved that:

the further consideration of the subject [of the bridge] be adjourned until the [board's] Engineer has conferred with the City

Architect and inspected the plans prepared by him, and has reported to the Committee upon such plans, and also generally his own views upon the subject.

The meeting between Sir Joseph and Horace Jones took place on 5 December 1877. Both men left the meeting with the distinct impression that the other was not really interested in any kind of collaboration. Sir Joseph was annoyed that Horace Jones was not prepared to allow him to formally assess his plans for a low-level bridge, preferring to consider them as not yet sufficiently well developed for rigorous scrutiny. Horace Jones was annoyed that the MBW was presuming to design a bridge for the Corporation. Perhaps he was also a little embarrassed that his plans were still at the conceptual level and a little bruised by the realisation that, as an architect, he was not really equipped to design such a technically challenging bridge.

Sir Joseph set about the completion of the only part of his brief that he could fulfil – that of presenting his own views on the subject of the new bridge.

A great man who made things happen, Sir Joseph already had three alternative designs for a high-level bridge on his drawing board. The bridge was to have a clear headway of 65ft above Trinity high water and a width of 60ft. The main carriageway would be 36ft wide and two 12ft wide footways were to be provided. The straight northern approach was to have

BELOW **Sir Horace Jones, by Walter William Ouless.** *(LMA/Collage 11920)*

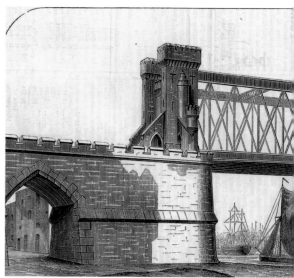

an incline of between 1 in 50 and 1 in 60. The southern approach was to be in the form of a spiral some 800ft in diameter with an incline of 1 in 40.

Sir Joseph was a visionary. He could foresee that sail would soon give way to steam and that a 65ft headway would be sufficient for all future needs. A high-level bridge would eliminate the need for expensive operating plant, and ongoing maintenance costs would be modest. His first and most straightforward design (below) was for a bridge of three simply supported steel lattice girder spans, resting on two piers in the river to provide a central channel of 444ft and two side channels of 184ft. The spacing of the piers took into consideration the tiers of ships maintained on the Middlesex and Surrey banks at that time. The girders were some 50ft tall, giving the bridge a workmanlike but austere appearance. There is no doubt that this was a practical design, but it was not a pretty one.

The second design (bottom) provided the same channels for shipping as the first solution but presented a more pleasing appearance. Each pier supported steel lattice cantilever sections. The shoreward sections were firmly secured to the bridge abutments, while the sections bearing the roadway towards the centre of the river carried a simply supported arched steel lattice girder. The whole is very reminiscent of the Firth of Forth Rail Bridge (but on a much smaller scale) on which construction would commence in 1882.

LEFT Bazalgette's high-level lattice girder bridge. *(ICE)*

LEFT Bazalgette's high-level cantilever bridge. *(ICE)*

The third Bazalgette design (above) was for a very elegant single-span braced-arch bridge crossing the river in one span of 850ft, with the arch having a rise of one-eighth of the span. Its appearance was not unlike the much later Tyne Bridge. It offered a 65ft headway for a channel width of approximately 580ft. Sir Joseph's estimate of the cost of this option was £1.25 million, plus £870,000 gross (£690,000 nett) for the approaches.

In his report, dated 22 March 1878, Sir Joseph described his third design only. He described the easy gradients, stressed the advantages to be enjoyed by vehicular traffic and foot passengers, and set out the reasons why his proposed design was to be preferred.

It seems that *The Engineer* was sent a copy of the report, as the issue of 29 March 1878 was full of praise for the single-span bridge:

Sir Joseph Bazalgette has proposed a scheme for getting rid of the London Bridge difficulty which is worthy of the metropolis, of the Thames, and of himself. Rising with the occasion, bearing in mind that he is dealing with the interests of an enormously wealthy city, and conscious of the vast constructive resources of the modern engineer, Sir Joseph proposes to construct the largest arch in the world, for such will be the great bridge of 850ft. span to be constructed, as we hope, across the Thames near the Tower. The scheme is sufficiently magnificent to startle those who are not well versed in modern bridge work. But to those who are, the only startling feature is that the proposal has a fair prospect of being carried into effect. When we remember the petty squabbling and the miserable contentions which have hitherto marked the dealings of the Metropolitan authorities with London Bridge, and the cheese-paring and unworthy proposals which have been made to patch and alter Rennie's great work, it seems almost incredible that a proposal should be made under the auspices of any Metropolitan body for the construction of what can hardly fail to be the finest bridge in the world.

Sir Joseph submitted his proposals to the MBW board without seeking the approval of the City Architect. There is no doubt that this was regarded as a snub by Horace Jones and the Corporation. MBW initiated the preparation of a Parliamentary Bill to build the high-level bridge. This was the second MBW initiative to be successfully opposed by the Corporation and the Thames Conservancy Board due to the restricted headway of the bridge. The headway of 65ft over Trinity high water was deemed inadequate, based on evidence presented by wharfingers to a parliamentary committee consulting on the Tower Bridge Bill, and members did not like the approach arrangements on the Surrey side.

In October 1878, Horace Jones released a report commenting on Bazalgette's designs. His view was that a high-level bridge of the headway proposed disregarded the evidence, views, wishes and interests of the wharfingers and shipowners trading between the proposed bridge and London Bridge.

Horace Jones's initial specification and outline design

Jones devoted the last quarter of his report to setting out and promoting his own proposals for the bridge. It should be a low-level bascule bridge which, when closed, should offer the same headway above Trinity high water level as London Bridge. The approaches would have easy gradients, minimising the amount of land which would have to be purchased. The centre opening would be 300ft – much wider than that possible with a swing bridge – and there would be less obstruction to navigation, with only two piers being required. The central span would comprise two hinged steel platforms, each lifted by eight chains passing over rotating barrels mounted in the arches to winding machinery

in the towers. A headway of 100ft would be provided over a channel width of 150ft. The side spans could be simply supported or supported by suspension chains. The architecture would blend with that of the Tower of London. His proposal is shown below, which is clearly the original inspiration for Tower Bridge as built, even if the towers have something of the fairytale castle about them.

After this promising start, it would seem that no concrete steps were taken to progress the project. The SBSC was dissolved in January 1879, steering of the project reverting to the BHEC.

Pressure from the public and businesses for the new crossing continued to build. Between 1874 and 1885 some 30 petitions from business houses, merchants and traders were presented to the Corporation, urging it to construct the new bridge – and quickly.

BELOW Horace Jones's first design for Tower Bridge. *(LMA COL 22427)*

Chapter Two

Tangible progress

Horace Jones's outline design for Tower Bridge becomes front runner in the informal design competition. Eminent engineer John Wolfe Barry is consulted and confirms that Jones' concept is realisable. The City's Court of Common Council backs the project and a Bill is prepared, receiving its third reading in July 1885. Jones is appointed architect and Barry engineer. Design work begins in earnest.

OPPOSITE The bascule bridges and the high-level walkways, with their views of London and their adrenalin-inducing glass floors, capture most of the limelight. But the suspension bridges are magnificent. The design of the main chains, which features twin catenaries, is inspired. Tower Bridge's possibly unique 'twin cable' design allows the bridge to cater for any load profile with minimal bending of the chains and roadway girders. *(Shutterstock)*

The impact of the Tay Bridge disaster

On Sunday 28 December 1879, shortly after 7.00pm and in extremely high winds, a large portion of the Tay Bridge fell into the water below. A train which had left Edinburgh at 4.15pm was on the bridge at the time and was precipitated into the stream. There were about 80 people on the train, all of whom lost their lives. This incident sent shock waves through both the Victorian engineering profession and the general public. All thoughts of bridge building were set aside pending the report of an inquiry. The report of 30 June 1880, concluded that 'The fall of the bridge was occasioned by the insufficiency of the cross bracing and its fastenings to sustain the force of the gale.'

The designer, Sir Thomas Bouch, had used a wind pressure of 10 pounds per square foot in his calculations. The disaster spelled the end of Bouch's career and he was replaced as engineer on the Forth Bridge in January 1881.

A committee was established by the Board of Trade, comprising eminent engineers of the day, including Sir John Hawkshaw and Sir William G. Armstrong, to set out five rules to be applied by bridge designers when performing wind loading calculations. For example, for railway bridges and viaducts, a maximum wind pressure of 56 pounds per square foot was to be used. This was the figure which would be applied for the new Tower Bridge when it came to be designed a few years later. The correspondence column in *The Engineer* in the months following the committee's report demonstrated that engineers felt that wind pressures above 40 pounds per square foot were unheard of in these islands and that the committee had erred very much on the side of caution.

In June 1881, the MBW offered an olive branch to the BHEC in the form of an invitation to a conference on the proposed Tower Bridge. This conference took place in the October of that year but did not result in the joint promotion of the project.

Strong protests about the delay in the provision of a new crossing were aired at a public meeting at the Mansion House on 25 May 1882, focusing the minds of the BHEC once again. During 1882, the BHEC considered three options: a proposal to free the Thames Steam Ferry from tolls; a proposal for a chain ferry; and the bridge proposed by Horace Jones. But no recommendation was made.

Last throw of the dice for Sir Joseph Bazalgette

Also in 1882, no doubt frustrated by the continued and palpable lack of progress on the project, Sir Joseph Bazalgette submitted a revised proposal for his braced arch high-level bridge, increasing the headway from 65 to 85ft above Trinity high water level and improving the southern approaches. This proposal carried a cost estimate of £1,845,000. He also submitted proposals for low-level bridges (with and without provision for opening). The estimate for the low-level bridge without an opening included provision for £2.25 million of compensation for wharfingers. This again illustrates his extraordinary foresight; in just a few years, entry of large ships into the Upper Pool would simply not be required. He also proposed a tunnel, as well as suggesting a project to widen London Bridge. This was to be the last throw of the dice for Sir Joseph.

It is worth noting that Sir Joseph's bridge proposals did not all fall upon stony ground. He designed and supervised the implementation in 1884 of modifications to the newly constructed but structurally unsound Albert Bridge, designed by Roland Mason Ordish; he designed and supervised the building of Putney Bridge in 1886, built by contractor John Waddell at a cost of £240,433, and he also designed and supervised the building of the new Hammersmith Bridge, replacing an elegant suspension bridge designed by William Tierney Clark which was proving incapable of meeting traffic demands. This was very similar to both the extant bridges over the Thames at Marlow and over the Danube at Budapest. The new Hammersmith Bridge was built by Dixon, Appleby & Thorne at a cost of £82,117. Two private Bills were introduced in early 1883, one for a duplex bridge and one for a subway – both private ventures seeking to earn revenue from tolls. These were successfully opposed by the Corporation. In June 1883, the Corporation consulted with the Thames Conservancy Board about a possible chain ferry at Greenwich; this came to nothing. Also in 1883, further consideration was given to the introduction of a free steam ferry.

1884 – decision time

In March 1884 there were three Bills before Parliament:

1 The Metropolitan Board of Works (Thames Crossing) Bill, which proposed a tunnel.
2 The (Duplex) Tower Bridge Bill, proposed by Bell and Miller, backed by private venture capital.
3 The Lower Thames Steam Ferries Bill, proposed by the Corporation.

A Select Committee of the House of Commons sat for 25 days, taking evidence from interested parties and reporting on 4 July that: 'Your Committee are of the opinion that two crossings are required and should be sanctioned by Parliament. The one a low-level pivot swing bridge at Little Tower Hill; the other a subway at or near Shadwell.' The committee expressed the hope that the Corporation would build the bridge and the Metropolitan Board of Works the tunnel.[1]

The Select Committee report seemed to provide the spur to action so desperately needed and, on 24 July 1884, the BHEC reported that the Corporation should construct an opening low-level bridge, funded by Bridge House Estates. The committee also offered to firm up the design and cost estimates with a view to an application being made to Parliament later in the year.

The Engineer of 1 August 1884 reported:

It is with some satisfaction that we draw the attention of our readers to the present condition of this scheme, and to the probability of a long-needed public improvement being carried into effect in the manner we have, in the face of much opposition, been recommending for the last ten years. From an engineer's point of view, there has never been any question that a low-level bridge near the Tower can alone

supply the accommodation needed. The beginning of the end seems to have come at last; the Committee of this session has not only recommended a low-level bridge with mechanical openings, but that the City and not the Metropolitan Board shall construct it. We hope no time will be lost, and that the Corporation, moribund though it may be, will prove equal to the occasion, and give us a structure worthy of the metropolis.

On 29 October 1884, BHEC presented a report to the Court of Common Council detailing the results of inspections made of low-level opening bridges in Great Britain and the Continent. The report contained sketches by Horace Jones of swing bridge and bascule bridge variants of the new Tower Bridge. Jones had also consulted John Wolfe Barry and a letter from Barry was appended to the committee's report, endorsing the practicability of the outline designs. Barry is believed to have spent a period of five weeks on this consultation, during which he prepared

BELOW **The Court of Common Council meeting in October 1884, by C.F. Kell.** *(LMA/Collage 3720)*

---

1 The tunnel – the Rotherhithe Tunnel – was opened in 1908, after much local opposition.

RIGHT John Wolfe Barry. *(Author/ICE)*

BELOW Early Jones/ Barry design. *(ICE)*

a cost estimate of £750,000 to build the bridge, which included the cost of land to be purchased to enable the project.

John Wolfe Barry was the son of Charles Barry, the celebrated architect who had designed the new Palace of Westminster. Barry was a skilled civil engineer and had been a pupil of John Hawkshaw, with whom he had designed and built the Cannon Street and Charing Cross railway bridges.

The Court of Common Council unanimously adopted the bascule bridge proposal. A Bill was to be prepared and considered by the House in the spring of 1885. An early Jones/ Barry design is shown here (below). Gone is the semi-circular arch, to be replaced by high-level walkways. The side spans are part cantilever and part suspension bridges. The abutments are of insignificant size and the piers are narrow.

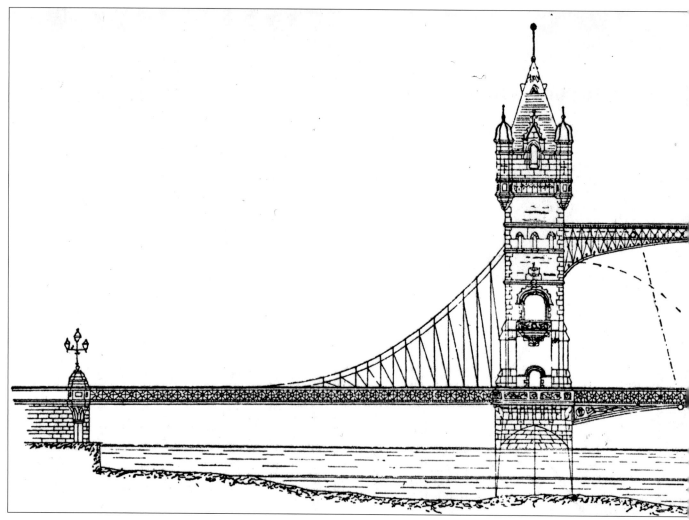

The Engineer of 7 November contained an editorial summarising the progress of the mooted bridge from 1866 to that time, but expressing concerns as to whether Horace Jones was the right man for the job:

With every respect for Mr. Horace Jones, the genial and popular City architect, we are bound to say that he is hardly the man to design the bridge, although he may protect the interests of the Corporation in determining on a design. If we may be allowed to say so, there is too much architecture and too little engineering in his scheme. Without in any way pinning him to what he frankly described as a preliminary sketch, we do not think that the maker of such a sketch can have a right conception of the work in hand. The shape and position of the curved chains, the combination of such chains and bracing with a lattice or trussed girder, are all ideas long ago exploded as an erroneous system of construction.

The editorial went on to suggest that the right way to take the project forward would be to issue plans and sections of the site and approaches and then to issue a formal invitation to professional engineers to submit their designs. These would be formally evaluated and a winning design selected. This did not happen.

Unsolicited designs kept coming, including one for a combined road and rail bridge by Roland Mason Ordish and Ewing Matheson. The bridge would start life as a bascule road bridge, with the rail bridge added once the Upper Pool had ceased to attract ocean-going vessels such that the openings could be permanently closed. *The Engineer* continued to rail against the Horace Jones design, but to no avail.

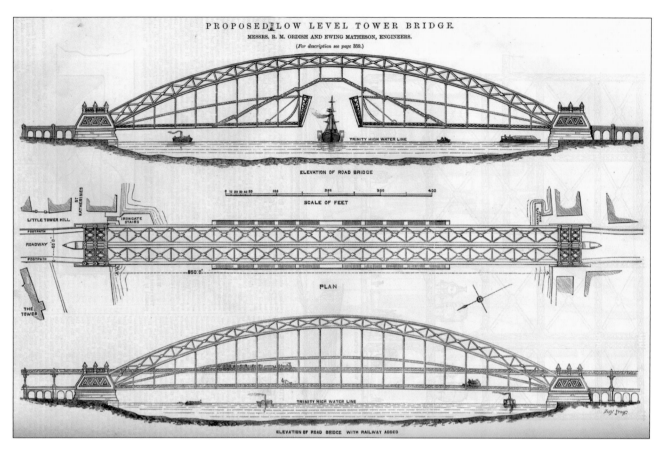

ABOVE Ordish and Matheson proposal for Tower Bridge. *(ICE)*

The die was cast and the Bill received its third reading in July despite strong opposition from wharfingers of the Upper Pool. Royal Assent was received on 14 August and in September the Court of Common Council authorised the BHEC to build the bridge. Horace Jones was appointed architect and John Wolfe Barry engineer. Jones was knighted in recognition of his services to architecture on 30 July 1886.

And so, a project first mooted in 1871 was finally up and running and in very good hands.

The detailed design phase

From the summer of 1885, Barry worked on the design with little input from Jones as the bridge was effectively a steel bridge with a stone façade. Barry had been in practice with Henry Marc Brunel (son of Isambard Kingdom Brunel) since 1878, and Brunel was given responsibility for much of the detailed design work and its attendant calculations. A Mr Fyson was employed by Barry to prepare detailed

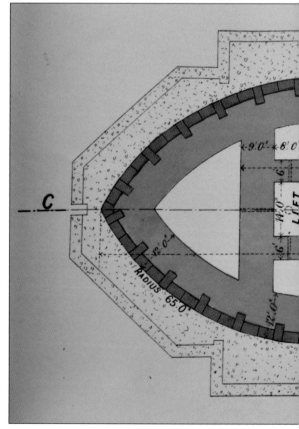

working drawings, and Mr George Daniel Stevenson took responsibility for the detailed architectural design of the external masonry. It is unclear whether Stevenson was employed by the City Architect's office or by Barry directly. Design and drawing work would occupy this team for the next four years.

Tenders for Contract No. 1 (piers and abutments) were delivered before noon on Monday 29 March 1886, so by that date Barry must have prepared at least an outline design of the whole bridge, sufficient to produce a close estimate of the total weight the foundations were required to bear, plus a detailed design and specification for the piers and abutments.

Study of the signed contract drawings for Contract No. 1 reveals that very detailed work had indeed been done. Highly detailed drawings of the permanent caissons, temporary caissons and end caissons were provided to the contractor, together with a drawing showing a suggested method for sinking the caissons and building the piers.

However, what is of particular interest is that the internal structure of the piers shown in

Contract Drawing No. 5 is very different from the piers as built. The drawings (on this page and overleaf) show chasing for opening chains, anchorages for the sheaves of the opening chains and two hydraulic lifts, presumably for the purpose of lifting the bascules. The accumulator chambers (if that is what they are) are sized differently, and there are no passages between the accumulator chambers and the bascule chamber. A detailed drawing is also provided of anchor plates and ties which are clearly not a part of the pier as built. It would appear that, in March 1886, John Wolfe Barry was still planning for the bascules to be opened by means of chains. This gives us a clear insight into why John Jackson's contract value increased by £10,000, namely some substantial changes were requested by Barry during the implementation of the contract and John Jackson was fully entitled to increase his price to accommodate them.

Work began without delay in April 1886, and a memorial stone was laid by HRH the Prince of Wales on 21 June 1886, with music supplied by the Band of the Coldstream Guards, at a cost

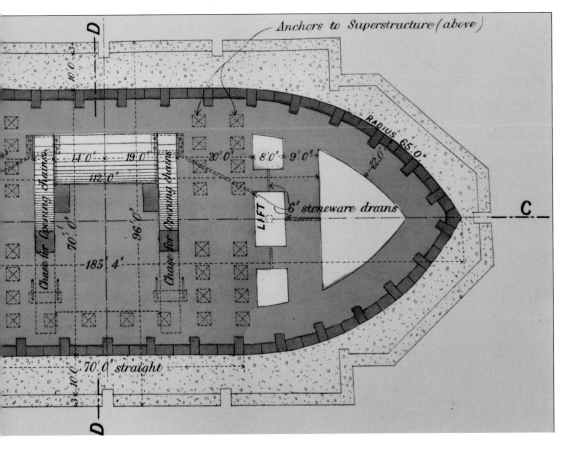

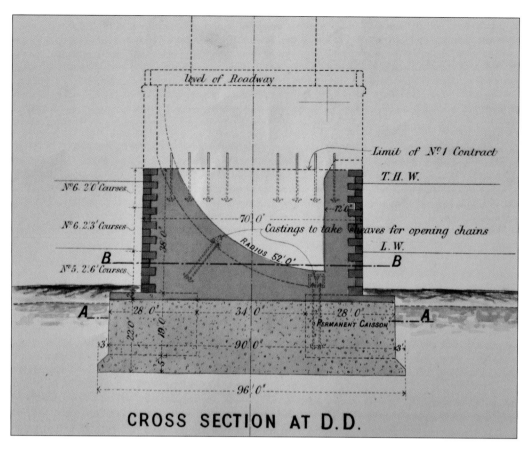

CROSS SECTION AT D.D.

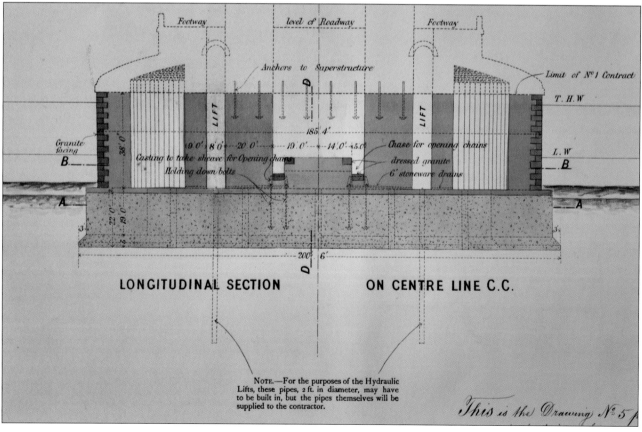

LONGITUDINAL SECTION   ON CENTRE LINE C.C.

NOTE.—For the purposes of the Hydraulic
Lifts, these pipes, 2 ft. in diameter, may have
to be built in, but the pipes themselves will be
supplied to the contractor.

This is the Drawing Nº 5

of £44 2s. It is the writer's belief that detailed design work on the superstructure of the bridge must have been ongoing throughout 1886, 1887 and 1888.

It is extraordinary that a bridge of such magnitude and complexity could be designed by such a small team, using drawing boards, slide rules and trigonometrical tables, and with no access to electronic calculators, spreadsheets, CAD packages, 3D finite element analysis software, copiers or printers.

Laying the memorial stone

The *London Gazette* of 29 June 1886 contains the loyal address given to HRH Albert Edward, Prince of Wales KG as follows:

We, the Lord Mayor, Aldermen, and Commons of the City of London in Common Council assembled, heartily and gratefully welcome the presence of Your Royal Highness, on behalf of Her Most Gracious Majesty the Queen, upon an occasion so interesting, and in its object so important, to the commercial interests of this vast Metropolis.

The Corporation of London has possessed for centuries estates charged with the maintenance of London Bridge. These estates were partly bestowed by generous citizens and partly derived from gifts made at the Chapel of Saint Thomas á Beckett, on London Bridge, for the maintenance of the bridge.

By careful husbanding and management of these estates, the Corporation has been enabled during the present century to rebuild, entirely free of cost to the ratepayers, London Bridge and Blackfriars Bridge, and to purchase and free from toll Southwark Bridge.

These obligations being provided for, the Committee charged with the management of the Bridge House Estates, brought up to the Court by the hand of their Chairman, Mr. Frank Green, in 1884, a full and exhaustive report, with plans, recommending that application be made to Parliament for powers to construct a new bridge across the River Thames from the Tower, which was agreed to.

In the session of 1885, the same

Committee, under the chairmanship of Mr. Thomas Beard, successfully promoted a Bill in Parliament, authorizing the construction of a Bridge, to inaugurate which, in the name of Her Majesty the Queen, your Royal Highness so graciously attends to-day.

Its completion within the space of four years, at a cost of 750,000 pounds, will supply a paramount need that has been sorely felt by dwellers and workers on the north and south sides of the Thames below London Bridge, and at the same time will greatly relieve the congested traffic across that ancient and famous thoroughfare.

In conclusion, we desire to express on this the first day of the fiftieth year of Her Majesty's happy and prosperous reign, our unswerving loyalty and devotion to Her Majesty the Queen, and to heartily thank Your Royal Highness for the important part you have been pleased to undertake in the great work before us, enhanced as it is, to our intense gratification, by the graceful presence of Her Royal Highness the Princess of Wales, to whom, with Your Royal Highness, we wish long life and all prosperity and happiness.

Perhaps the inclusion of the budgeted cost and planned time-scale of the bridge was a mistake!

His Royal Highness gave the following answer:

Gentlemen, it gives the Princess of Wales and myself sincere pleasure to be permitted on behalf of The Queen, my dear mother, to drive the first pile of the New Tower Bridge, and in Her name We thank you for your loyal Address, and assure you of Her interest in this great undertaking.

All must allow that this work, when completed, will be one of great public utility and general convenience, as tending materially to relieve the congested traffic across this noble river.

We shall always retain in Our remembrance this important ceremony.

We cordially thank you for the very hearty welcome which you have accorded to Us, and We will not fail to communicate to The Queen the sentiments of affectionate attachment which you have expressed.

The bridge as built

The image below shows the final design. John Wolfe Barry describes it in a paper which is included in a history of the bridge written in 1894 by Charles Welch, librarian to the Corporation of London. Let's hear the description in Barry's own words:

The problem to be solved in the design of the Tower Bridge was one of no small difficulty, for it was necessary to reconcile the requirements of the land traffic with the very important interests of the trade of the Upper Pool. This part of the river is always crowded with craft of various kinds, and it was this fact that made the 'bascule' system so desirable. Any opening bridge revolving horizontally would have occupied so large an area of the river as to be very undesirable from many important points of view, whereas a bridge revolving in a vertical plane, not only occupies the minimum of space in the river, but also at an early stage of the process of opening affords a clear passage for ships in the central part of the waterway, increasing in width rapidly as the operation of opening is continued.

The mode in which the traffic of the Pool is conducted prescribed the general arrangement of the spans of the bridge. Sea-

going vessels of all kinds are moored head and stern in two parallel lines in the Upper Pool, on each side of the centre line of the river, leaving a central channel from 200 to 250 feet wide free for the passage of vessels up and down the river, and this space is frequently contracted by barges and small craft lying alongside the larger vessels, to a width of from 160 to 180 feet. The spaces in the river occupied by the large vessels on each side of the free central channel are called tiers, and as vessels lie in the tiers two or sometimes three abreast, with barges alongside them, it will be seen that if the piers of a bridge were made alignable with the tiers there would be no obstruction to navigation, and little to the flow of water, by two piers of a width not greater than that of the tiers. On the landward side of each of the tiers, channels are preserved for the passage of vessels to and from the wharves, and it was of course necessary that these side channels should not be obstructed by any pier of the bridge. Thus the mode in which the river traffic has for many years adjusted itself, made it evident that a bridge with a clear central opening of from 160 to 200 feet, and two side openings of about 280 or 300 feet, would meet all requirements, and that there could be no objection to piers

wide enough to accommodate a counter-balance, seeing that the width of two vessels lying in the tiers would be more than the width necessary for such an extension of the moving girders into the piers as would provide for a sufficient counterpoise.

With these few words on the principles that governed the main features of the Tower Bridge, we will proceed to consider the details of the structure generally.

The Act of Parliament prescribed the leading dimensions of the Tower Bridge to be as follows:

1   A central opening span of 200 feet clear width, with a height of 135 feet above Trinity high water when open for vessels with high masts, and a height of 29 feet when closed. (It may be mentioned in passing that the central span has been made 6 inches higher when closed than was stipulated, and is, as executed, of the same height as the centre arch of London Bridge which is 29 feet above Trinity high water, and that it is 5 feet higher when open than was prescribed by Parliament).
2   The size of the piers to be 185 feet in length and 70 feet in width.
3   The length of each of the two side spans to be 270 feet in the clear.

The mode adopted for spanning the landward openings is by suspension chains, which, in this case, are stiffened. The chains are anchored in the ground at each end of the bridge, and are united by horizontal ties across the central opening at a high level. These ties are carried by two narrow bridges 10 feet in width, which are available as foot bridges when the bascule span is open for the passage of vessels. The foot bridges are 140 feet above Trinity high water, and as their supports stand back 15 feet from the face of the piers, their clear span is 230 feet. Access is given to them by hydraulic lifts and by commodious staircases in the towers.

Above the landings at the tops of the stairs, and on which the foot passengers land from the lifts, come the roofs of the towers, the crestings on the tops of which are 206 feet above

the roadway level, or 298 feet from the bottom of the foundations.

The fixed parts of the superstructure of the Tower Bridge consist, as has been said, of two shore spans, each of 270 feet, and of a central high-level span of 230 feet. The fixed bridge is of the suspension form of construction, and the chains are carried on lofty towers on each pier and on lower towers on each abutment.

It was originally intended by Sir Horace Jones, the City architect, that the towers should be of brickwork in a feudal style of architecture and that the bridge should be raised and lowered by chains somewhat like the drawbridge of a Crusader's castle. Subsequently, Sir Horace Jones proposed a combination of brick and stone.

The ideas were in this condition when the writer was appointed engineer to the scheme, with Sir Horace Jones as architect, and the Corporation went to Parliament for powers to make the bridge. It was seen that any arched form of construction across a span to be used by masted ships was inadmissible, and that whatever headway was given should be absolutely free of obstruction throughout the whole width of the span.

Sir Horace Jones unfortunately died in 1887, when the foundations had not made much progress, and up to that time none of the architectural designs had proceeded further than such sketches and studies as were barely sufficient to enable an approximate estimate to be made of the cost. Since the death of Sir Horace Jones, the general architectural features of the Parliamentary sketch designs have been preserved, but it will be seen that the structure as erected differs largely therefrom, both in treatment and material.[2] The width, and consequently the weight, of the bridge was increased by the requirements of Parliament, and the span of the central opening was enlarged from 160 feet, as originally intended, to 200 feet. At the same time the provision of lifts and stairs to accommodate foot passengers when the bridge was open was felt to be a necessity.

2   Mr George D. Stevenson took on responsibility for the detailed architectural treatment.

In this way it became apparent that it would not be possible to support the weight of the bridge on towers wholly of masonry, as in the first designs, unless they were made of great size and unnecessary weight. It was consequently requisite that the main supports should be of iron or steel, which could, however, be surrounded by masonry, so as to retain the architectural character of the whole structure.

The skeleton of each tower consists of four wrought steel pillars, octagonal in plan, built up of riveted plates. The pillars start from wide spreading bases, and extend upwards to the suspension chains, which they support. They are united by horizontal girders and many diagonal bracings. The chains are carried on the abutments by similar but lower pillars.

On the tops of the octagonal pillars rest a series of rollers which will allow the chains so to move as to accommodate themselves to changes of temperature and to unequal distribution of the road traffic.

The ends of the chains on the abutments and on the towers are united by large pins to the ties. The ties on the abutments are carried down into the ground below the approaches, and are there united to anchorage girders, which rest against very heavy blocks of concrete, and are abundantly adequate to resist the pull of the chains. The ties at the high level between the towers are for the purpose of uniting the upper ends of the two chains, and by this means the stress on the chains is conveyed from anchorage to anchorage.

Ten bridges in one

Tower Bridge as built comprises ten separate bridges.

There are two suspension bridges, one on either side of the river linked by central ties, the dead weight of which is carried by the high-level footbridges. The suspension bridges use rigid steel girders as 'chains' rather than the wrought-iron links used on other bridges of the period or the wire ropes used on the Brooklyn Bridge and later suspension bridges. The individual girders and ties are linked by large steel pins. Steel rockers are provided to

transfer load to the steel columns in the main towers and abutment towers. Land ties connect the rockers in the abutment towers to large fabricated steel anchor girders embedded in concrete deep below ground.

There are two simply supported, fabricated steel beam bridges carrying the roadway over the bascule chambers and supporting the weight of the bascules.

There are arguably *two* bascule bridges (many bascule bridges have just one bascule).

There are two steel cantilever girder bridges built out across the river from each of the main towers. Two simply supported girder bridges are suspended between them. These structures form the high-level walkways and carry the dead load of the central ties of the suspension bridges by means of vertical rods suspended from the upper boom of the outer girder of each footbridge. I consider the completed high-level walkways to be two bridges.

There are two masonry arch viaducts, with accommodation under the arches, forming the approaches of the bridge (a section of each approach is also carried between plain masonry retaining walls).

On the northern side, guard house and storeroom accommodation was provided for the Tower of London; on the southern side, engine rooms, boiler rooms and a coal store were provided to house the machinery required to generate the hydraulic pressure needed to operate the bascule bridges and control systems. The Guard House under the northern approach is now a restaurant, but accommodation under other arches is still used by the Tower of London.

Now twelve

In 1960, two further suspension bridges were added to relieve the cantilever bridge walkways from the load of the high-level ties. Each 'bridge' comprises two 60mm-diameter galvanised locked-coil wire ropes anchored to the steel superstructure of each tower. New suspenders transferred the dead weight of each tie, which is some 80 tons (not including the weight of the eyes at the ends), to the cables. Modern sources tell us that the cables were each tensioned to 100 tonnes.

Chapter Three

A bridge is born

Tower Bridge used cutting edge technology. No one company possessed the expertise to build the bridge on a turnkey basis, so the work was divided into eight fixed-price contracts, let to five contractors. Contract terms were onerous (liquidated damages and a 10% retention of contract value refundable on completion) and timescales were tight. Close cooperation between contractors was demanded in the contracts.

OPPOSITE Tower Bridge. *(Shutterstock)*

The method of contracting

The Corporation of London was accustomed to designing and specifying its new bridges using its own City Architect, an appointed external engineer and its own teams of engineers and technicians. The work would be broken down into the minimum number of separate contracts and, for each, a detailed contract document would be prepared, containing a specification of works, a bill of quantities and a schedule of prices to be completed by the tenderer for the pricing of each element of the work. One set of contract drawings would accompany each contract.

These contracts were very professionally drafted and very nicely presented, being typeset and bound. The contract drawings were beautifully drawn by hand and colour washes applied. Contractors were responsible for preparing any tracings they required over and above the single copy of each relevant drawing provided. There is no doubt that John Wolfe Barry (the engineer) and his expert team were largely responsible for the technical content in the contract documents, which was detailed, extensive and extremely professional in its execution.

For each contract, advertisements would be placed in the press inviting interested parties to tender. Tenders would be opened at a meeting of the Bridge House Estates Committee and the successful tenderer selected.

This was the method adopted for Tower Bridge and it was probably the only method possible, no single company being capable of undertaking the wide range of specialised work required to build the bridge on a turnkey basis.

Having said that, no evidence of newspaper advertisements for Contract No. 5 has been found, either in newspaper archives or in the journal of advertisements maintained by the Comptroller of Bridge House Estates. This points clearly to Contract No. 5 being awarded on a single tender basis to Sir W.G. Armstrong, Mitchell & Co. Ltd. For Contract No. 5, there really was only one company in Britain with the skill and experience to design and build the hydraulic machinery for the bridge. This contract was let in December 1887, before the design work on the hydraulics of the bridge had been undertaken.

There is no evidence in the ledgers of a separate design contract, so it is likely that Armstrongs were required to estimate the cost of design work up front, estimate the size of machinery required for the bridge and the sum total of pipework required *before* the design work had been done, and provide a fixed price covering the design work and the supply and installation of all of the steam and hydraulic equipment.

Many drawings were produced by Armstrongs during the 21-month period from contract signature to 30 September 1889, the date of the first payment of £20,000 made to the company. These include the drawing shown on page 75, which is dated 23 January 1889. There is evidence of design and drawing work being undertaken by Armstrongs from contract start to a month after the bridge opened.

The individual contracts

The work was broken down into the following contracts, the table opposite showing the name of the successful tenderer in each case.

John Wolfe Barry's estimate of £750,000 for the bridge included a sum of £165,000 for land purchase, so the £830,005 building cost represents an increase above the estimated building cost of 42%.

A further contract, for the construction of a water reservoir, was let to John Mowlem. The reservoir was for the purpose of storing, for reuse, the water exhausted from the hydraulic engines.

The contracts followed a logical sequence. The piers and abutments had to be largely in place before the contract could be let for the masonry over about 4ft above Trinity high water. The contract for the hydraulic machinery of the bridge had to be let some six months before the contract for the southern approach, as the basic layout and design of the engine room accommodation had to be determined before contract drawings for Contract No. 4 could be completed. Contracts 6 and 7 had to start together as these contractors would need to work closely together to complete the masonry of the piers to carry the steel girders supporting the roadway over the bascule chambers and the granite slabs and holding-down bolts to secure the steel columns.

The contracts made explicit provision for this

Contract No.	Scope	Contractor	Date of contract	Contract value
1	Piers and abutments	John Jackson	April 1886	£131,344[1]
2	Northern approach, including north anchorage girders	John Jackson	March 1887	£52,882
3	Cast-iron parapet for northern approach	John Jackson	December 1887	£5,596
4	Southern approach, including south anchorage girders	William Webster	July 1888	£38,383
5	Hydraulic machinery	Sir W.G. Armstrong, Mitchell & Co. Ltd	December 1887	£85,232
6	Iron and steel superstructure	Sir William Arrol & Co. Ltd	May 1889	£337,113
7	Masonry superstructure	Messrs Perry & Co., (Herbert Henry Bartlett)	May 1889	£149,122
8	Paving and lighting	Messrs Perry & Co.	May 1889	£30,333
				£830,005

1   The contract signed on 13 April 1886 was for £121,344.

concurrency. For example, Contract No. 6 for the superstructure (iron and steel work) contains the following wording:

12. The Corporation have already let the following Contracts:

No. 1. – For the construction and maintenance for twelve months after completion of the two Piers in the River, and the two Abutments up to the level of 4-feet, or thereabouts, above Trinity high water, and the providing of the anchor ties and holding down bolts required for the machinery or other parts of the Superstructure within and about the Piers and Abutments.

No. 2. – For the construction and maintenance for twelve months after completion of the Northern Approach and the completion of the Northern Abutment up to the level of about 23-feet above Trinity high water. This Contract also includes the Northern Anchorage of the Bridge.

No. 3. – For the supply and erection, and maintenance for twelve months after completion, of the cast-iron parapets for the Northern Approach.

No. 4. – For the construction, and maintenance for twelve months after completion, of the Southern Approach, and the completion of the Southern Abutment up to the level of about 20-feet above Trinity high water. This Contract also includes the Southern Anchorage of the Bridge.

No. 5. – For the supply and erection and maintenance for two years after completion, of the Hydraulic Machinery for working the Bridge.

Contracts Nos. 1, 2, and 3 are let to Mr. John Jackson; No. 4 to Mr William Webster; and No. 5 to Messrs. Sir W. G. Armstrong, Mitchell & Co., Limited.

12A. The Corporation intend to let at the same time with this Contract, a Contract (No. 7) for the masonry, brickwork, &c., for the completion of the Piers and Abutments up to the level of the bottom of the pillars supporting the Superstructure, and for the completion of the Towers on the Abutments and on the Piers in the River. A condition will be inserted in that Contract that the Contractor therefore shall so arrange the execution of the work included therein as to prepare in a convenient and expeditious way for the erection of the ironwork and steelwork of the Superstructure, and to afford all facilities for that purpose. The Contractor under this Contract, will be similarly bound to afford all due facilities for the execution of the Contracts Nos. 5 and 7, and for the maintenance of the work included in Contract Nos. 1, 2 and 4.

However, it cannot have been easy for Arrol's and Bartlett's men to erect the steelwork of the main towers and the masonry 'cladding' of the towers at the same time.

The tendering process

As an illustration of the tendering process, the Notice for Contract No. 1 reads as follows:

TOWER BRIDGE. – CONTRACT No. 1.

Notice is hereby given, that the Bridge House Estates Committee of the Corporation of London will meet at Guildhall on Monday, the 29th day of March next to receive TENDERS for the CONSTRUCTION of the LOWER PORTIONS of the ABUTMENTS and TWO PIERS of the proposed TOWER BRIDGE.

The Specification of the Works, Form of Contract, Form of Tender, Bill of Quantities, together with the Contract Drawings, may, from and after the 15th March next, be seen at the Architect's Office, Guildhall, E.C., and at the office of John Wolfe Barry, Esq., No. 23, Delahay-street, Westminster, and copies of all the above documents may be there obtained on payment of five guineas, which will be returned to all unsuccessful competitors who send in a bona fide Tender and return all the documents.

The Tenders must be on the printed Form of Tender, which must not be detached from the Specification and Quantities, and they must be sealed up, addressed to the Comptroller of the Bridge House Estates, Guildhall, London, E.C., endorsed "Tender for the Tower Bridge – Contract No. 1," and be left at his office before Twelve at noon, on Monday, the 29th March, and the parties tendering will be expected to be in attendance at Guildhall on that day, at One o'clock precisely.

The Committee do not bind themselves to accept the lowest or any Tender that may be sent in.

JOHN A. BRAND
Comptroller of the Bridge House Estates.

This notice was published in *The Times*, *The Standard*, *The Daily News* and *The Morning Post* on 1, 3, 5, 8, 10, 12, 15, 17, 19 and 22 March 1886. It was published in *The Telegraph* and *The Advertiser* on 2, 4, 6, 9, 11, 13, 16, 18, 20 and 23 March. It also appeared in *The City Press*, *The Citizen*, *The Metropolitan*, *The Builder*, *The Engineer*, *The Architect*, *The Building News* and *British Architect*.

So bidders had just two weeks to prepare their tenders for this very large fixed-price contract, which featured a liquidated damages clause and extremely tight timescales. They were not asked for a written proposal, setting out their understanding of the requirements, their approach to the work, a detailed project plan and work breakdown structure, curricula vitae of the people who would manage the project, the company's previous experience of work of this type, method statements and a risk register. All the Corporation required was a completed schedule of prices.

The committee would meet on the day the tenders were delivered, with the clear expectation that a supplier would be selected on that day, with contract signature following just a few days later. There could be no evaluation of proposals, as no proposals had been requested. It is not clear whether John Wolfe Barry was even present when the tenders were opened. It is also unclear from the notice whether the contractors were expected, or would be permitted, to make a presentation to the committee. It is unlikely.

Contract No. 1

To illustrate the look and feel of a Tower Bridge contract, and provide evidence of the care with which it was drafted to cover every aspect of the work required, while ensuring that as much risk as possible was borne by the contractor rather than the City, a small extract from Contract No. 1 (piers and abutments) is included opposite.

The contract clearly sets out the anticipated timescale. Work had to begin within one week of receipt of award of contract. One pier and its adjoining abutment had to be delivered within 15 months of contract start, with the other pier and its abutment being delivered

THE TOWER BRIDGE.

CONTRACT No.1.

SPECIFICATION OF WORKS

To be executed in the construction, completion and maintenance of certain parts, hereinafter more particularly described, of a Bridge across the River Thames, adjoining the Tower of London, situate in the Counties of Middlesex and Surrey, and authorised by 'The Corporation of London (Tower Bridge) Act, 1885,' (hereinafter referred to as 'The Tower Bridge Act,') and other works connected therewith.

I. – EXTENT OF CONTRACT.

1.   The Contract based on this Specification is made with the Mayor and Commonalty and Citizens of the City of London.

2.   The Contract provides for the construction and completion in accordance with the requirements of the Tower Bridge Act, and the maintenance for twelve months after completion, of the two abutments and the two piers of the Tower Bridge from their foundations up to a height of four feet or thereabouts above the level of Trinity High Water, and the providing and fixing of all such anchors, ties, holding-down bolts or other ironwork which may be required for the machinery or other parts of the superstructure of the Bridge, within or about the said abutments and piers, together with all other works connected with the said abutments and piers, as authorised by the Tower Bridge Act, and in accordance with the Contract Drawings and with this Specification.

3.   The Contract comprises the providing of the necessary materials of every kind, labour of every description, pumps, dredgers, grabs, machinery, mortar mills, engines, boilers, pile-driving machinery, temporary roads, piling, platforms and staging, all scaffolding, tackle, cranes, steam tugs, punts, barges, rafts, carts and other means of transport, and all plant, tools, implements, and every appliance, whether temporary or permanent, necessary for the due and satisfactory execution and completion of the works above described, and for the maintenance thereof, and also the providing and erecting such places as may be necessary for depositing and protecting materials, machines, articles and things aforesaid to the satisfaction of the Engineer.

4.   The whole of the materials and workmanship are to be of the very best description, and the whole of the works are to be completed to the entire satisfaction of the Engineer. All ironwork and steelwork, must, unless specially ordered by the Engineer, be manufactured within the United Kingdom.

5.   The works included in the Contract comprise the fencing (whether temporary or permanent) of the properties required for the abutments on either side of the River, and all lands which the Corporation may see fit, on the application of the contractor, temporarily to occupy for the purpose of the said works; the erection of the cofferdams requisite for and the excavating of the foundations of the abutments; the erection of the stages in the River surrounding the sites of the two piers, together with all fenders, booms, dolphins and other machinery, implements and appliances necessary for the same, and for the guidance and protection of the traffic on the River, as may be from time to time required by the Thames

Conservancy, or as are required by the Tower Bridge Act; the providing the ironwork and steel for the caissons, and the erecting, timbering, and sinking the same; the excavations of the abutments and piers; the filling in of the excavated material where required, and the removal of the surplus excavation; the building of all concrete brickwork, or masonry, and the providing and fixing of all ironwork used in the construction of the abutments and piers; and, generally, the construction and completion of all works described or shown in the Contract Drawings, or described, or shown, or marked in any writing or detailed drawing which may from time to time be furnished by the Engineer, explaining the details of the said works or the method of executing the same; and all other works properly incidental to the performance and completion of the Contract, whether specifically described or not, which may be ordered from time to time by the Engineer.

II. – LIST OF CONTRACT DRAWINGS ACCOMPANYING THIS SPECIFICATION.

6.   The Contract Drawings referred to in this Specification are as follow:-

No. of Drawing
          1. – General Plan and Section.
   2, 3 & 4. – The Caissons for the Piers.
          5. – The Piers.
          6. – The Abutments.
          7. – Anchor Plates and Anchor ties in Piers.

7.   Besides the above a Drawing (No. 5A) has been prepared which does not form one of the Contract Drawings, but it may be examined by the Contractor in order that he may see the way in which it is suggested that the sinking of the caissons and the building of the piers may be done. The precise mode in which these operations are to be carried out may be altered by the Contractor, subject, however, to his submitting his proposed mode of operations for the approval of the Engineer, whose approval, nevertheless, will not (nor will the adoption of the mode shown on Drawing No. 5A, and described in this Specification, clauses 164 to 185) relieve the Contractor from his responsibility for the safe and efficient carrying out and completion of the works, nor shall such approval in any way be deemed a warranty that such mode of operation will be successful.

III. – GENERAL CONDITIONS AND STIPULATIONS.

8.   In this Specification the words 'Tower Bridge Act' shall be held to mean the Corporation of London (Tower Bridge) Act, 1885, and all Acts incorporated therewith.

9.   The word 'Corporation' shall be held to mean the Mayor and Commonalty and Citizens of the City of London, or any Committee or person duly authorised by them to act in their behalf.

10.   The word 'Contractor' shall be held to mean the person or persons whose tender for the work referred to in the Specification may be accepted by the Corporation, and shall include the heirs, executors, administrators, and permitted assigns of the said contractor.

11.   The word 'Engineer' shall be held to mean Mr. Horace Jones, Architect to the Corporation, and Mr. John Wolfe Barry, or either of them, or other the Engineer for the time being appointed by the Corporation; and the word 'Assistant' shall be held to mean any person or persons appointed by the said Engineer as his Assistant or Assistants in the work.

no later than 22 months from contract start. (The reason for the different dates was that the contractor was required to make a minimum width of channel available to river traffic, which restricted the staging that could be erected around the pier sites.) These timescales are extremely aggressive for work of this difficulty and structures of this magnitude. A liquidated damages clause was included, whereby the contractor was required to pay the Corporation £50 for every day that the contract was late, over £6,000 per day at today's prices.

The contract was signed on 13 April 1886.

John Jackson was also required to provide two sureties prepared to bind themselves in the sum of £10,000 each for the due performance of the contract. Jackson offered Alderman Preston, mayor of Sunderland, and Charles Jackson Esq of Newcastle upon Tyne as his sureties.

Completion of Contract No. 1 was in January 1890, which means that this part of the Tower Bridge project took almost exactly twice as long as planned. On 28 March 1893, George Cruttwell presented a paper on the foundations and piers of the Tower Bridge at the Institution of Civil Engineers. In the ensuing discussions, John Wolfe Barry expressed his view that John Jackson's price had been rather too low. He also mentioned that John Jackson had confided in him that he only made a small profit on the job.

Other contracts also specified demanding timescales. For example, Sir William Arrol was required to complete the superstructure within 24 months from receipt of order, while having his hands expertly tied by the following wording in his contract:

It is expected with regard to Contracts Nos. 1 and 2 that the Northern Abutment and the Northern Pier will be ready to be delivered to the Contractor at or about August next; the Northern Approach will be sufficiently advanced by about June next, and the Southern Pier by about December next. With regards to Contract No. 4, the Southern Abutment will be ready to be handed over to the Contractor by January 1890, and the Southern Approach by March 1890. In giving these dates it is to be expressly understood

that the Corporation does not bind itself to the accuracy thereof, and that no inaccuracy shall relieve the Contractor from his obligations under the Contract, or give him any right to compensation.

Project management

John Wolfe Barry could not have been on site at all times. He appointed Mr George Edward Wilson Cruttwell MICE, who had joined his practice in 1879, to act as resident/site engineer throughout the period taken to build the bridge. It seems very likely that Cruttwell was the 'Assistant' referred to in Contract No. 1.

The Engineer of 30 October 1885 ran this typically polemical piece:

THE TOWER BRIDGE
The last phase in this scheme, which recently received the sanction of Parliament, is somewhat ludicrous. It will be remembered that the design of Mr. Horace Jones, the City Architect, was adopted, a gentleman who had no experience in bridges, the Corporation jealously confining the matter to their own officer, and neither inviting nor allowing any suggestions from the outside. This course being also followed when the Bill came before Parliament, no alternative bridge design could even be considered, and the monstrous structure, about which we have already expressed our opinion, was authorised. Now, however, when the question of the architect's remuneration comes before the Court of Common Council, the same Bridge Committee which before Parliament said they had 'full confidence in their own officer' announce, what everyone outside the Court knew long ago, that the designing of bridges was not a proper function of the architect, that therefore his ordinary salary of £2,000 a year does not cover this particular matter, and that he must be paid sufficient to depute the work to someone else. The sum of £30,000, recommended as the remuneration for conducting so important a work, namely, about 5 per cent on the estimated expenditure, would not be excessive if it were to be paid to anyone but a salaried

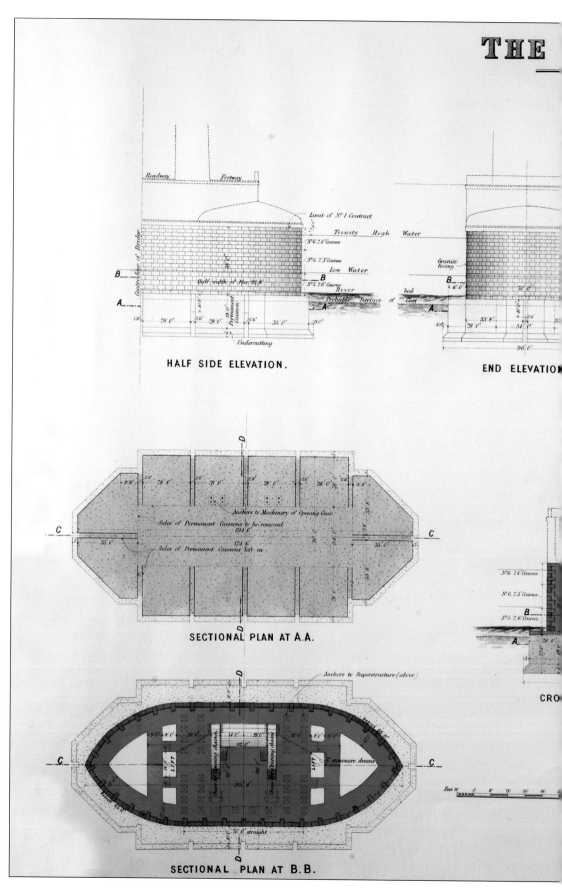

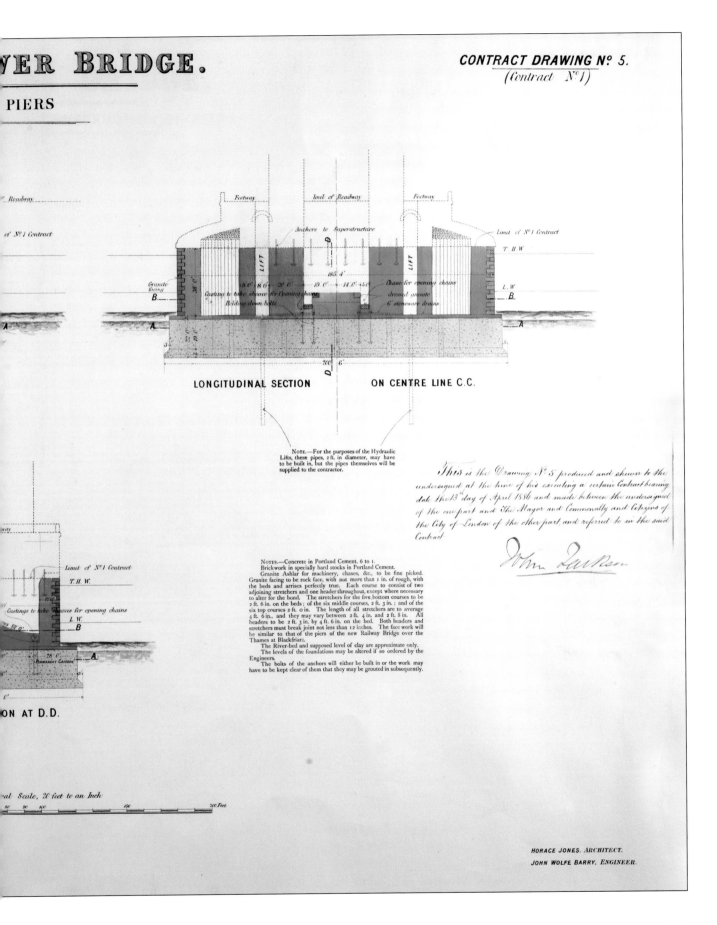

LONGITUDINAL SECTION   ON CENTRE LINE C.C.

NOTE.—For the purposes of the Hydraulic Lifts, these pipes, 2 ft. in diameter, may have to be built in, but the pipes themselves will be supplied to the contractor.

This is the Drawing № 5 produced and shewn to the undersigned at the time of his executing a certain Contract bearing date the 13th day of April 1886 and made between the undersigned of the one part and The Mayor and Commonalty and Citizens of the City of London of the other part and referred to in the said Contract

John Jackson

ON AT D.D.

NOTES.—Concrete in Portland Cement, 6 to 1.
Brickwork in specially hard stocks in Portland Cement.
Granite Ashlar for machinery, chases, &c., to be fine picked. Granite facing to be rock face, with not more than 1 in. of rough, with the beds and arrises perfectly true. Each course to consist of two adjoining stretchers and one header throughout, except where necessary to alter for the bond. The stretchers for the five bottom courses to be 2 ft. 6 in. on the beds; of the six middle courses, 2 ft. 3 in.; and of the six top courses 2 ft. 0 in. The length of all stretchers are to average 4 ft. 6 in., and they may vary between 2 ft. 4 in. and 2 ft. 8 in. All headers to be 2 ft. 3 in. by 4 ft. 6 in. on the bed. Both headers and stretchers must break joint not less than 12 inches. The face work will be similar to that of the piers of the new Railway Bridge over the Thames at Blackfriars.
The River-bed and supposed level of clay are approximate only.
The levels of the foundations may be altered if so ordered by the Engineers.
The bolts of the anchors will either be built in or the work may have to be kept clear of them that they may be grouted in subsequently.

Scale, 20 feet to an Inch

HORACE JONES, ARCHITECT.
JOHN WOLFE BARRY, ENGINEER.

officer; but instead of the money being paid to the engineer, over whom the Corporation would then have direct control, the amount is to be divided as the architect and engineer may agree.

The ledgers of expenditure on the Tower Bridge project show that, under an agreement made on 17 December 1885, a joint quarterly payment of £1,750 was indeed made to Horace Jones and John Wolfe Barry, the first payment being made on 27 July 1886. There is no indication as to how this payment was split between them, but in the absence of any other payments made to John Wolfe Barry or his engineering practice, it must be assumed that this payment was made largely to Barry's firm to fund the design work and the services of George Cruttwell. £1,750 per quarter for the four years planned for the project, plus a final payment of £2,000 does add up to £30,000.

The payment of £1,750 made on 30 July 1887 was to the executors of Sir Horace's estate. After the death of Sir Horace, payments of £1,670 per quarter were made directly to Barry, the final payment of £2,000 being made on 6 June 1893.

Tower Bridge contracts required the contractors to provide fortnightly returns. For example, on Contract No. 1, John Jackson was required to report the number of men employed by type and the number of horses working at the site.

The ledgers reveal that all contractors received regular payments on account, the sums often being in round figures.

At the end of every month, each contractor was required to review the work done during the month with the engineer or, more likely, with resident engineer George Cruttwell, to ascertain by actual measurement the quantity of work completed. The Engineer would then issue a certificate, permitting the Corporation to pay the contractor for 90% of the value of work done. The 10% retention on total contract value was not released until contract completion was signed off by the Engineer.

Project plans maintained by contractors would, in all probability, have been handwritten.

BELOW Part of the Tower Bridge progress record showing the main towers, the bascules and the high-level walkways. It provides an invaluable record of the progress of the build. *(LMA COL/ PL/01/192/B/006)*

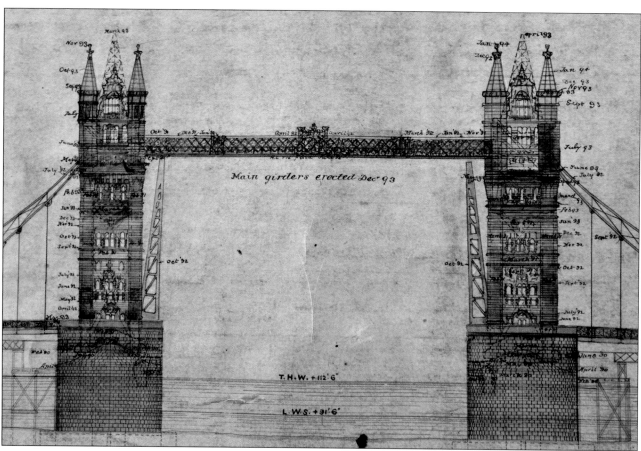

Barry's team maintained a detailed record of progress. For Contract No. 1, a progress record was maintained for each pier, by caisson number. These records show that for Caisson 13 (on the Surrey pier):

Caisson arrived	22 August 1887
Commenced erecting Caisson	26 August 1887
Commenced lowering Caisson	15 September 1887
Caisson at bed of river at +78' 6"	16 September 1887
Commenced grabbing	8 October 1887
Pumped out first time at +64' 9"	31 October 1887
Pumped out last time at +63' 0"	2 November 1887
Caisson at bottom at +59' 6"	5 November 1887
Finished undercutting	14 November 1887
Finished concrete	24 November 1887
Brickwork at +78' 6"	30 May 1888
Commenced setting granite	5 June 1888

The Surrey pier progress record also shows that excavation of the silt from the space inside the ring of 12 caissons commenced on 16 September 1889 and was complete on 14 November 1889. Concrete laying was completed on 25 November 1889 and brickwork was finished up to +116ft 6in (the limit of Contract No. 1) on 11 February 1890.

Barry's team also maintained progress records for the blockwork on each face of each pier, the records showing when each section of granite blocks were in place.

A wonderful survivor is a progress record (see opposite page and below) on which the completion date of each principal component of the bridge was noted. It is a lovely thing, at least 2m long and covering the entire length of both approaches. It is part detailed drawing and part watercolour. It looks very much like the work of George Stevenson and would probably have hung on the office wall of George Cruttwell, John Wolfe Barry or George Stevenson.

BELOW Part of the Tower Bridge progress record that includes the northern abutment and the anchor ties. It is believed to be the only publicly available drawing that shows the location of the pinned joints linking the Middlesex land ties to their anchor girders. *(LMA COL/ PL/01/192/B/006)*

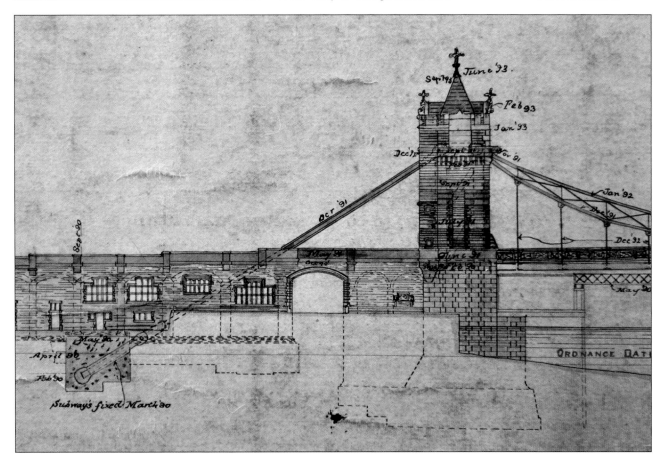

Construction of the piers

There is an extremely detailed account of the method of constructing the piers in the paper George Cruttwell presented to the Institution of Civil Engineers on 28 March 1893 (Paper No. 2652). A summary will suffice here.

The drawing on pages 52–53 includes a 'Sectional Plan at A.A.', which shows eight rectangular caissons and four caissons of an approximately triangular shape in plan. A caisson is basically a box, with removable sides, constructed of wrought iron and strutted with sturdy timbers to withstand water pressure. A rolled steel cutting edge was riveted to the bottom edges of each caisson. The caissons were erected on temporary wooden staging and, after the river bed had been levelled by divers, sunk into the bed of the river by applying kentledge (heavy weights, typically made from pig iron).

The lowest section of each caisson, which would not be removable after the pier had been completed, was known as permanent caisson; these sections of caisson were 19ft high. Once the permanent caissons were settled on to the river bed and commencing their downward journey, further sections of temporary caisson

were bolted to them, the joints being sealed with India rubber.

Divers worked inside each individual caisson to excavate first gravel and then London clay. When the permanent caisson had penetrated about 5ft into the clay, the water in the caisson could be pumped out, allowing navvies to replace divers on excavation duty at periods of low tide. Work continued 24/7. Additional lengths of temporary caisson were added as each caisson sank, until the bottom of each permanent caisson was 20ft below the river bed. The water could then be excluded permanently. The clay was then excavated 7ft deeper than the bottom edge of the caisson and outwards beyond the cutting edge for a distance of 5ft on three of the four sides of the caisson. By this means the whole foundation was made continuous. Concrete gauged at six parts of Thames ballast to one part of Portland cement was then laid up to a level about 6in above the cutting edges, preventing further movement. The sides of the permanent caissons facing the centre of each pier were prepared for removal by cutting out a number of rivets and installing wooden boards to prevent concrete from adhering to the steel wall, and wooden boxes were installed against the inside surface of these sides to form dovetails. The permanent caissons and the spaces between them were then filled with concrete, enabling the inner sides of the permanent caissons to be removed a few days later.

Timber piles were driven between the caissons, registering in grooves near the outer corners of the caissons. The joints between the piles and the caissons were rendered watertight, allowing the sides of the temporary caissons to be removed. This enabled the construction of a continuous masonry wall within the temporary caissons which was both watertight and able to resist the pressure of the water when the outer sides of the temporary caissons were later removed.

The next step was the excavation of the space between the two rows of four rectangular caissons and the filling of this volume with concrete and brickwork. An engraving of the piers under construction can be seen on the opposite page.

Actual timescale and cost

The bridge took eight years to construct, twice the originally planned timescale. With hindsight, the original expectation must be seen as overly ambitious.

With regard to cost, a paper presented to the Institution of Civil Engineers on 10 November 1896 by George Cruttwell states that the total cost was £902,500, excluding the cost of the land. Modern sources state the total cost as £1,184,000. The ledgers reveal a slightly different story. From 14 August 1885 (the date o f the Tower Bridge Act) to 31 December 1898 the total expenditure was £1,642,714 1s 5d. The reason for including the years ending 31 December 1895–98 is that substantial capital expenditure was still being incurred on the bridge, including a payment of £30,000 to Perry & Co. in 1898 in settlement of a claim.

This expenditure does include the purchase of land and buildings to enable the construction, the cost of compensation to wharfingers and others for deleterious impact on their businesses, plus legal costs and all ancillary costs associated with the project. The total cost of the project at 2019 prices is approximately £200 million. This only takes into account the change in the value of the pound, not the increase in labour rates since the 1890s. The Daywork Rates set out in Contract No. 6 make for interesting reading.

Hourly daywork rates, Contract No. 6	£	s	d
Cart, horse and driver	0	1	4
Cart, two horses and driver	0	2	0
Boy	0	0	4
Ordinary labourer	0	0	7
Excavator, pile-driver, timber-man, concrete-mixer, or watchman	0	0	9
Carpenter or joiner	0	0	11
Bricklayer	0	1	0
Mason	0	1	0
Smith	0	0	11
Fitter	0	1	0
Diver, signalman, two labourers to air-pump, and use of diving apparatus	0	15	0
Ganger	0	1	2
Foreman	0	1	8

There was no minimum wage in the 1890s! A boy working a 60-hour week would be charged at £1, the boy probably taking home considerably less to his mum.

Londoners were enthralled by the sight of the steel skeleton of the bridge taking shape.

The bascules were lowered for the first time on 27 March 1894, just three months before the opening ceremony.

RIGHT The Surrey main tower rising from its pier, sometime in 1891. *(LMA COL 323335)*

BELOW The cantilevered footbridges approaching each other in March 1892. *(LMA COL 323354)*

RIGHT This fascinating image is the only one the author has been able to find that shows the location of the lower pinned joint in a Middlesex land tie. Published in June 1892 in *The Graphic*, the image, which was probably engraved from a sketch rather than a photograph, shows the whole length of the land tie from its connection to the west Middlesex horizontal link to the lower pinned joint. The box girder section above the parapet is yet to have the top plates fitted, and the mating section of anchor tie, attached to the anchor girder, is nowhere to be seen.
(Look and Learn)

BELOW Taken on 24 September 1892, this photograph shows the main chains of the Surrey suspension bridge under construction.
(LMA COL 323356)

BELOW RIGHT Depicting progress at some time in 1893, the steelwork is complete, with the exception of the bascules, and the masonry cladding has reached the third level.
(LMA COL 169070)

THE CORPORATION OF THE CITY OF LONDON.

Opening of THE TOWER BRIDGE, June 30

BY

H.R.H. THE PRINCE OF WALES. K.G.

ON BEHALF OF

HER MAJESTY THE QUEEN.

THE RIGHT HON.ble GEORGE ROBERT TYLER. Lord

JOHN VOCE MOORE ESQre ALDERMAN. } Sheriffs.
JOSEPH COCKFIELD DIMSDALE ESQre ALDERMAN. }

ADMIT   David Thomas Esq

Albert J. Alb

DOMINE   DIRIGE   NOS.

HONI·SOIT·QUI·MAL·Y·PENSE

GRACE

DURUM PATIENTIA   FRANGO.

SOLO   DEO   SALUS.

SC 131/188

Blades, East & Blades, 23, Abchurch Lane, London.

The grand opening

The bridge was opened on 30 June 1894 by the Prince of Wales on behalf of Her Majesty Queen Victoria. A few minutes before the ceremony, John Wolfe Barry was advised that he had been made a Companion of the Order of the Bath.

The weather was splendid, with a wonderfully clear atmosphere and a sky of peerless blue. This contrasted with the day eight years earlier, when the memorial stone had been laid, on which the weather had been foul. The event drew massive crowds. At 11.50am, the royal carriage arrived at the steps of the Mansion House bearing HRH the Prince of Wales, the Princess of Wales, the Duke of York and Princess Victoria of Wales. The royal party was greeted by the Lord Mayor, the Right Hon George Robert Tyler and the Lady Mayoress. After a brief welcoming speech, the Lord Mayor, sheriffs and other civic dignitaries entered their carriages and preceded the royal procession to the bridge.

A mighty cheer and the sound of drums precisely at noon heralded the approach of the royal party to the thousands of seat-holders stationed in specially constructed grandstands on both sides of the northern approach. The band of the Coldstream Guards was playing and, as soon as the procession was visible from the river, steamships, tugs and launches sounded their whistles, sirens and bells. The procession drove over the bridge to the Surrey side, turned around and returned to the Middlesex side where the royal party entered the profusely decorated royal pavilion.

After a welcoming address by the Recorder, Sir Charles Hall, the Prince of Wales gave a short speech hailing the 'splendid engineering skill bestowed upon the construction of the bridge'. John Wolfe Barry then advanced to the dais and shook hands with the prince, who announced 'I declare this bridge open for land traffic.'

The prince then turned to a pedestal on which was a silver gilt loving cup. Removing the cover of the cup and applying it to a switch device, the prince twisted the cover to send a signal to the Surrey engine driver, who lifted his bascule, the Middlesex driver following suit. The royal party left their places to watch the bascules

open. The prince's announcement, 'I declare this bridge open for river traffic', was drowned out by cheers from the crowds, a further sounding of whistles and sirens by river craft, a flourish of trumpets and a royal salute from the guns of the Tower of London. A flotilla of 12 decorated vessels passed through the bridge.

The prince was then presented with the silver gilt cup and the princess with a diamond pendant/brooch. The royal party then proceeded to the Tower for a state visit. One imagines that a sumptuous lunch, together with liquid refreshment, was provided for the many dignitaries in attendance.

The first year of operation

The bridge was originally required to stay in the bascules-raised position for two hours around the time of high tide. This was very unpopular with road users and pedestrians and the practice was soon discontinued. By the end of its first year of operation, the bridge was carrying one-third of the traffic that had been seen on London Bridge prior to the opening. This was despite the fact that the London County Council (LCC) had delayed the linking of the southern approach to New Kent Road and Old Kent Road. LCC took over the responsibilities of the Metropolitan Board of Works in 1889. The delay was caused by LCC's unsuccessful attempts to persuade Parliament to agree to a scheme whereby property owners would be required to contribute to the cost of the works in return for the increased value of their properties which would result.

River traffic was similarly heavy, the bascules being raised 6,160 times. Ships wishing to pass through the bridge were required to signal their intention by suspending two black balls from a mast. (Perhaps fluorescent orange would have been a better colour.) The City was obliged to provide the services of a tug to assist vessels passing through the bridge, this responsibility lasting until after the Second World War. Vessels are sometimes still towed through the bridge today.

Personnel

Some 82 staff were employed to operate the bridge. These included:

- The Tower Bridge Master, for whom a residence was provided in the southern abutment tower.
- The Deputy Tower Bridge Master, for whom a residence was provided in the northern abutment tower.
- The Tower Bridge Engineer.
- A superintendent of machinery.
- Stokers and oilers to work in the plant rooms and the machinery rooms on the piers.
- Watchmen, signalmen, gatemen and relief men to manage the processes established to safely manage the lifting of the bridge when required.
- Drivers to raise and lower the bascules.
- Liftmen, painters, electricians, gas and water fitters, clerks and messengers.

Today, the bridge is maintained and operated by a technical staff of around 15 people, but many more than this are employed to run the bridge as a business and an attraction.

BELOW The achingly pretty PS *Waverley* being towed through the bridge on 5 October 2018. She looks a little ashamed, but it was a wise move. Paddle steamers do not respond well to the helm and have an enormous turning circle. She would never have been able to turn in the Inner Pool under her own steam. *(Author)*

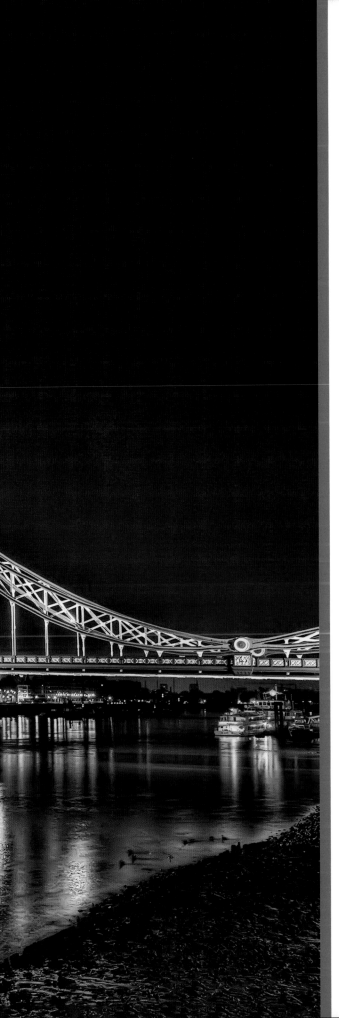

Chapter Four

The suspension bridges

The elegant chains of the suspension bridges are beautifully illuminated at night. But much impressive technology is hidden from view. This includes the rockers which originally allowed the bridges to adjust to temperature change and live loading. Plus the huge concrete anchor blocks buried deep beneath the approach viaducts, which resist the tension in the main chains and ties.

OPPOSITE In this stunning view of Tower Bridge all lit up at night, the suspension bridge chains are clearly illuminated. *(Shutterstock)*

Sources

Two sources in particular shed light on the design of the suspension bridges. George Cruttwell was John Wolfe Barry's site engineer. He presented two papers to the Institution of Civil Engineers, one on the foundations of the river piers and one on the superstructure of the bridge. These appear in the proceedings of the institution.

James Tuit was Sir William Arrol's design engineer. He wrote a very substantial paper for *The Engineer* covering the design and construction of the bridge, later published in book form. These sources contain much useful information and many drawings. Unfortunately, few of the drawings in these three papers are of sufficiently high quality for publication, which is why photographs are provided of parts of the bridge structure wherever possible.

Overview

On each suspension bridge, the load of the roadway is suspended from four 'chains', two on either side of the roadway, the chains taking the form of braced steel girders which are stiff enough to prevent local deflection of the roadway under traffic loading. The chains are unequal in length, this design decision being taken in the light of the different heights of the main towers and the abutment towers. The main chains are spaced 18.4m (60ft 6in) apart from centre to centre.

For each suspension bridge, the distance in plan between the points in the main tower and its adjoining abutment tower from which the chains are suspended is some 91.7m (301ft). Each suspension bridge carries the load of its suspended roadway by means of 28 vertical steel suspender rods (14 on each side of the roadway), plus the two suspension points under the pins linking the chains.

The main towers comprise steel superstructures, the weight and imposed loading of which is carried on masonry piers which are massive in themselves and which also bear the applied load of the bascule bridges, the road bridges which span the bascule chambers, the high-level footbridges and the masonry which surrounds the steel towers.

The steel superstructure of each tower, which is of all-riveted construction, comprises four octagonal steel columns, one in each corner. The columns are rigidly attached to three substantial landings, and provided

BELOW This elevation/section of the bridge (Fig. 1 of Cruttwell's paper) shows that the width spanned by each suspension bridge, measured between the face of the river pier and the face of the adjoining abutment is 82.3m (270ft). This is also the navigable channel width on the Surrey side, the channel width being some 3.6m narrower on the Middlesex side. The headway available to river traffic above Trinity high water level is 6.1m (20ft) next to the Surrey abutment, 7.0m (23ft) next to the Middlesex abutment and 8.2m (27ft) next to the piers. *(ICE)*

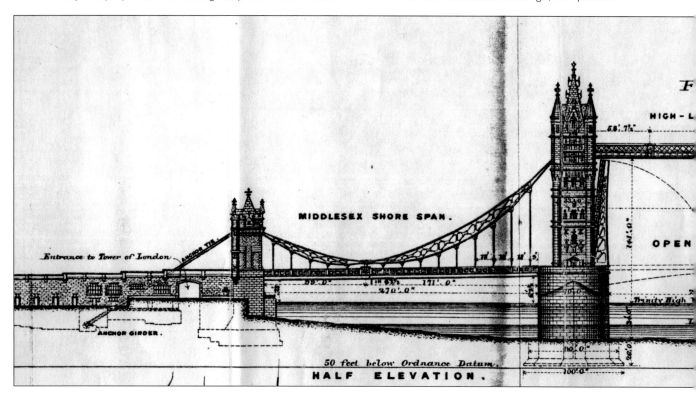

with cross bracing to withstand wind loading transmitted from the external masonry 'cladding'. The steel roof structure of each tower is supported by a further landing which is mounted on four steel stanchions positioned above the columns.

Each abutment tower also comprises a steel superstructure and an outer masonry skin, the weight of which is carried on a substantial brick pier sitting on a concrete foundation. The steel superstructure of each abutment tower also comprises four octagonal steel columns, one in each corner, stiffened longitudinally by horizontal and diagonal girders and transversely by arched steel box girders above the roadway.

Below the approaches at each end of Tower Bridge are two massive steel anchor girders set deep underground and with a great mass of concrete holding them in place. These four anchor girders resist the tension in the chains and ties induced by the suspended load of the two suspension bridges.

Moving from right to left in the elevational drawing below, and considering just the linked members on the upstream (west) side of the bridge, the tensile load is first carried by the west Surrey land tie which is a riveted steel girder some 51m (167ft) long, taking the form

of a box girder above the parapet and a pair of flat ties of rectangular section, each comprising a number of thick plates riveted together, below ground. As the west Surrey land tie approaches the west Surrey anchor girder, it branches to provide three points of attachment to the anchor girder, the attachment being by means of riveting.

At the landward side of the Surrey abutment tower, the west Surrey land tie connects via a pinned joint to a horizontal link comprising two parallel thick steel plates made up of multiple plates riveted together. The river end of the link is pinned to the west Surrey short chain. Below the two pinned joints, the link is supported on rollers (or 'rockers'), which transmit vertical load to the two westward octagonal columns of the Surrey abutment superstructure and provide the means whereby movement caused by bridge loading and thermal expansion or contraction is accommodated.

The next pinned joint links the long and short west Surrey chains above the suspended roadway.

The west Surrey long chain is pinned to the west high-level tie above the west landward column of the Surrey main tower. The end of the tie is supported on rockers which transmit

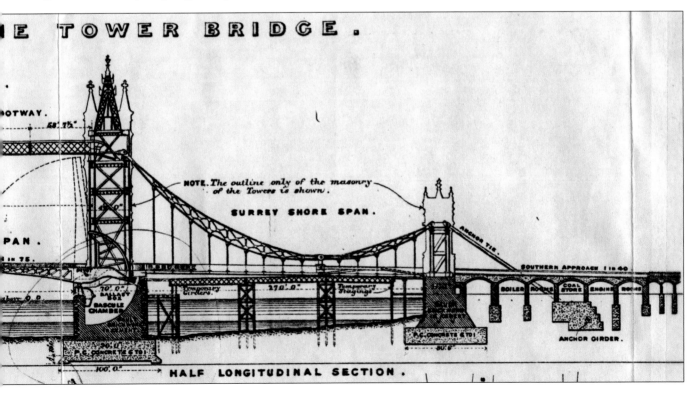

ABOVE **Scrap of the Ordnance Survey map published 1880–82, showing the site of the northern approach.** *(National Library of Scotland)*

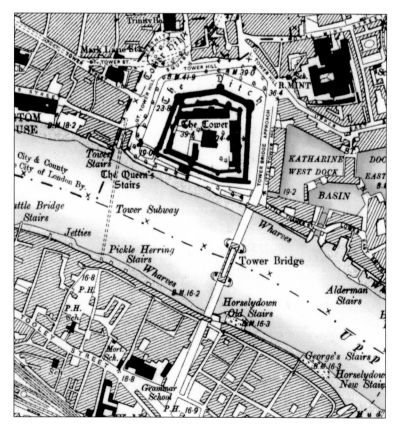

the vertical load to the west landward column of the Surrey main tower.

The west high-level tie carries the tensile load to the Middlesex main tower. The arrangement is similar on the Middlesex side of the bridge. However, the form of the Middlesex land ties below the bridge parapet differs from that on the Surrey side. The means of connection of the Middlesex land ties to their anchor girders is also different.

The chains and ties were tensioned simply by removing the temporary staging and support girders below the suspended roadway sections when bridge construction was complete.

The northern approach, a masonry arch viaduct

Preparation

It was necessary to undertake a grisly task before work could commence on the northern approach. The Corporation of London issued the following advertisement in *The Times* and five other newspapers on 18 December 1886:

The Corporation of London under the powers and provisions of the Corporation of London (Tower Bridge) Act 1885 propose to remove the remains of the several persons interred in the Burial Ground situate [sic] at the East end of the Tower of London to the City of London Cemetery at Ilford. Notice is hereby given that the heirs executors administrators or relatives or friends of any person whose remains are so interred in the said Burial Ground may in accordance with Section 40 of the above Act remove the remains of such person to any other Burial Ground wherein burials may legally take place upon making due application at the

LEFT **This is a scrap of the Ordnance Survey map published in 1894–96. Comparison with the OS map from 1880–82 (above) shows that the northern approach stole some ground from the eastern ditch of the Tower and a lot of ground from the existing Little Tower Hill Road, which was narrowed to around one-half of its original width.** *(National Library of Scotland)*

*Comptroller's Office Guildhall on or before
the 20th day of January next at which Office
every information will be afforded.*

*Dated this 18th day of December 1886
John A Brand, Comptroller of the Bridge
House Estates, Guildhall E.C.*

This burial ground, which occupied half of the
eastern ditch of the Tower, is clearly visible
on the Ordnance Survey map published in
1880–82 seen opposite (top).

The site

The masonry arch viaduct, constructed of
brick piers and arches and faced with Kentish
ragstone on the western side only, represents
less than 40% of the northern approach. The
remainder is an elevated roadway supported
between brick retaining walls. The red brick
retaining walls on the western side of the
northern approach look a little tawdry compared
with the stone facing and one wonders why
this economy was made and how it would
have been acceptable to the Tower authorities.
However, the image of the viaduct section
facing the Tower makes it clear that the project
delivered a good deal of spacious and secure
accommodation for the Tower, including a
guard room next to a wide archway just to
the north of the abutment, which provided a
replacement eastern entrance to the Tower.
The existing guard room, visible in plan in the
scrap of the OS map shown opposite (top), was
clearly demolished.

Contract No. 2

The contract for the northern approach to
Tower Bridge, Contract No. 2, was let to John
Jackson on 28 March 1887. The contract value
was £52,882 0s 10d. No advertisement has
been found in either *The Times* archive or the
book of advertisements for tenders maintained
by the Comptroller of Bridge House Estates,
so it is likely that this contract was let to John
Jackson on a single-tender basis.

Either way, the contract gave Jackson
just one year from contract start to build
the northern approach up to the level of the
underside of the paving of the approach, and
this at a time when he was just 12 months into
Contract No. 1, which would not be complete

until January 1890. The work included the
continuation of the northern abutment, faced
with granite, up to the level of the underside of
the paving, unless the Corporation decided to
let Contracts 6 and 7 before Contract No. 2
was complete, in which case the abutment
was to be built up to a level 10ft lower. The
work included the provision and fixing of anchor
girders within their massive anchor block, and
the diversion and reconstruction of sewers,
pipes, and gas and water mains. The cast-
iron parapets and lighting were excluded from
the contract but were shown in the contract
drawings for completeness.

**ABOVE The northern
approach, viewed
from the north-west.**
(Author)

**BELOW The stone-
clad viaduct facing the
Tower.** *(Author)*

RIGHT Contract
Drawing No. 4, signed
by John Jackson. The
guard room is on the
left, the archway to the
Tower in the middle,
and the location of
the memorial stone,
still there today albeit
behind iron railings, on
the right. *(LMA)*

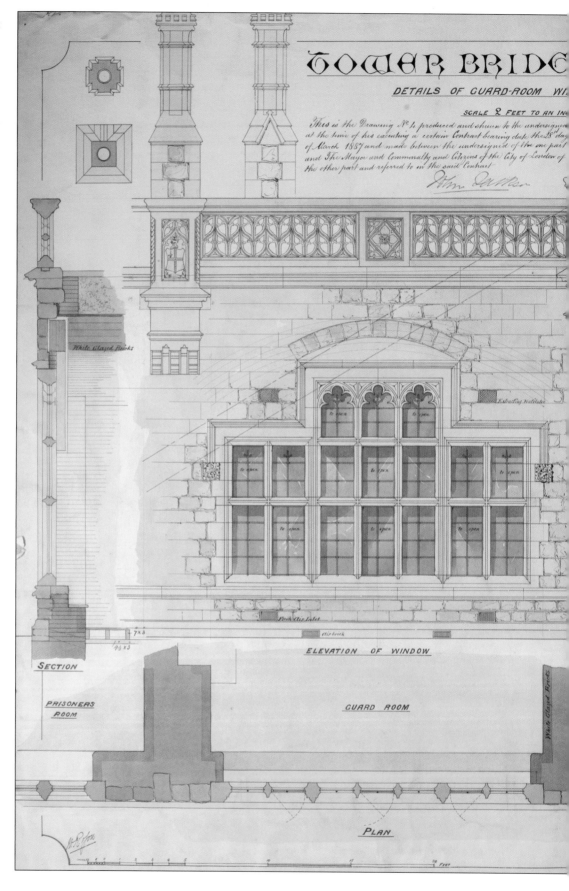

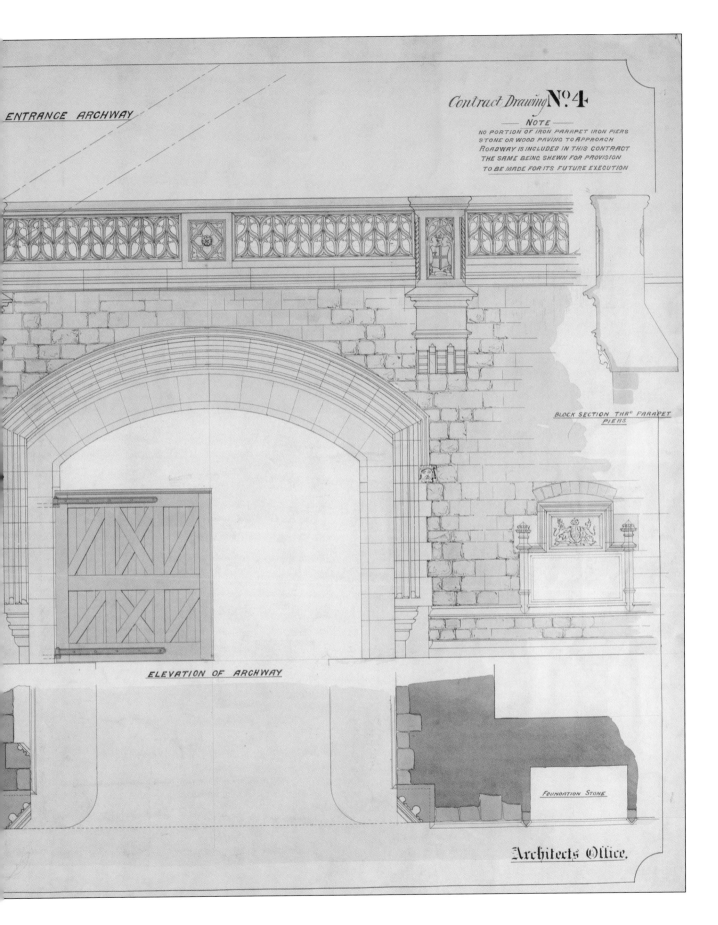

ENTRANCE ARCHWAY

Contract Drawing **№ 4**

NOTE

NO PORTION OF IRON PARAPET IRON PIERS
STONE OR WOOD PAVING TO APPROACH
ROADWAY IS INCLUDED IN THIS CONTRACT
THE SAME BEING SHEWN FOR PROVISION
TO BE MADE FOR ITS FUTURE EXECUTION

BLOCK SECTION THRO PARAPET
PIERS

ELEVATION OF ARCHWAY

FOUNDATION STONE

Architects Office.

ABOVE The view
from the guard room
window today, which
has been sensitively
converted into a
smaller window with
a doorway on either
side. *(Author)*

Accommodation beneath the arches of the viaduct

John Wolfe Barry tells us that accommodation was provided for both the bridge and the Tower. Clearly, the bridge engineers and technicians would need access to the anchor girders for inspection and maintenance purposes. The lower photograph on page 69 suggests that

vaults beneath 13 of the brick arches were used to provide accommodation, the three nearest the bridge being provided with attractive and large windows.

Today the guard room is a restaurant. The old prisoners' room is now the kitchen of the restaurant and toilets occupy what was once the small arms store.

RIGHT The chimney
of the guard room.
(Author)

The guard room would have been snug in winter. The chimney provided was designed to be as inconspicuous as possible among the row of ornate cast-iron gas lamps lighting the bridge – a change from the stone-built chimney depicted in Contract Drawing No. 4.

The northern abutment

The northern abutment features a large vaulted passage beneath, and parallel to, the roadway above, providing access to a landing stage and stairs. The stairs replaced the pre-existing Irongate Stairs, which Samuel Pepys records he used to take his most treasured goods, by lighter, to the south bank of the Thames on 4 September 1666, to escape the Great Fire. The iron gate from which the steps take their name was a large gate providing access to the Tower from the east.

This part of the northern abutment became known as 'Dead Man's Hole' as river currents had a tendency to cause bodies to be washed up there.

ABOVE '**Dead Man's Hole**': the room in the abutment is believed to have been used as a mortuary, but it is not known whether this accommodation was originally intended for this purpose. However, the white glazed bricks would be consistent with this usage, and these bricks are specifically mentioned in the Bill of Quantities in Contract No. 2. What *is* known is that the body of Josef Jakobs, a German spy executed by firing squad at the Tower on 15 August 1941, was brought here en route to its final resting place at Kensal Green Catholic Cemetery. *(Author)*

RIGHT **The east side of the Middlesex abutment tower taken from Tower Pier.** *(Author)*

Metrics

The contract required the excavation of 25,800yd^3 of material, the laying of 13,853yd^3 of concrete, and the setting of 14,693yd^3 of brickwork. Some 8,000ft^3 of Portland stone, 7,300ft^2 of Kentish ragstone facing, and 6,110ft^3 of granite was to be procured and fixed in place; 132.5 tons of iron and steel was required, which included the anchor girders and a large quantity of 12in iron pipe for conveying hot water from Aldgate station.

The southern approach, also a masonry arch viaduct

The site

Examination of 'before' and 'after' Ordnance Survey maps shows that a substantial amount of property to the west of Horselydown Lane had to be purchased from private owners to provide the necessary land for the southern approach.

Like the northern approach, the southern approach is part masonry viaduct and part elevated road supported by retaining walls. The masonry arch viaduct comprises a roadway arch and five vaults, constructed to house the original Tower Bridge machinery for generating the hydraulic pressure to operate the bridge. The retaining walls were constructed to provide cellarage for houses to be built later on either side of the new approach.

Contract No. 4

The contract for the southern approach to Tower Bridge was advertised in July 1888, with drawings, specifications and the Bill of Quantities being available to view from 11 July. Tenders were required to be delivered to the Comptroller of Bridge House Estates before 12 noon on Wednesday 25 July 1888. William Webster was the successful tenderer, probably signing the contract in August 1888. The contract value was £38,383.

Accommodation beneath the arches of the viaduct

This accommodation for bridge operating plant is described fully in Chapter 6. However, it is interesting to see from the plan below that some modifications to the initial layout

Outline arrangement of steam engines, boilers, accumulators pipes &etc
Scale ½ inch to 1 foot

LEFT An early plan of the plant rooms under the southern approach. *(Science Museum Archives MSP/0140/05)*

of the plant rooms and to the position of the accumulator tower were made during execution of the contract. Perhaps it was realised that an internal coal store would prevent pilferage and that substantial man effort would be saved if the coal store were to be positioned as close to the wharf as possible and right next to the boilers. Or did they just forget to include stairs for access to the plant spaces and subways below the floor?

The piers and their foundations

The force the London clay beneath each pier is required to support is approximately 6.97×10^5kN (70,000 tons). This comprises the weight of the concrete foundation; the masonry in the pier; the steelwork, iron and lead in the structure of the tower and the bascule; the imposed loading on the steel superstructure from the suspension chains, high-level ties and footbridges; and the weight of the masonry cladding of the tower.

On 28 March 1893, George Cruttwell presented his paper on the foundations of the river piers of the Tower Bridge to a meeting at the Institution of Civil Engineers. In the discussion which followed, John Wolfe Barry stated that, in the design of Charing Cross Bridge, he had used a maximum pressure on the clay of about 7 tons per square foot. For Cannon Street Bridge, the design pressure used was between 4.5 and 5 tons per square foot. In both those bridges, perceptible, but not serious, subsidence had occurred. He had been anxious to avoid this at Tower Bridge, given the lofty nature of the bridge.

George Cruttwell then described an experiment which had been done at the site of the bridge to determine what pressure on the London clay should be used for design purposes. A cylinder, 6ft in diameter had been sunk into the bed of the river and progressively loaded. At a pressure of 6.5 tons per square foot, the cylinder had begun to settle and the settlement had continued to increase. As a result, the decision was made to limit the design pressure on the clay to 4 tons per square foot (429kN/m^2). This determined that the required area of the foundation concrete would be $70,000 \div 4 = 17,500$ square feet ($1,626$m^2).

The length at the bottom of the foundation of each pier is 62.3m (204ft 6in). The maximum width is 30.48m (100ft). The height from the bottom of the foundation to the level of the roadway and footway is 28.5m (93.7ft). Standing at the bottom of the bascule chamber, one gets the feeling that one is deep below ground; the reality is that there is 11.7m (38.3ft) of brickwork and concrete beneath one's feet.

There was some discussion at the meeting mentioned above that the 'foot' around the bottom of the mass concrete foundation of each pier might be a point of weakness in that the concrete at the 'heel' of the foot might be in tension and therefore prone to fracture. Barry's view was that the foot could only be in tension if the concrete as a whole was seen as capable of bending; he did not feel that this could be the case.

The concrete foundation of each pier contains no steel reinforcement. It was poured as separate sections which are dovetailed together. The masonry above the concrete foundation up to a level of 4ft above Trinity high water was built by John Jackson under Tower Bridge Contract No. 1. It was mostly built using wire-cut gault bricks (smooth, heavy, yellow clay bricks) using 2.5:1 sand to Portland cement mortar. The facing brickwork in the bascule chambers and accumulator chambers was built using Staffordshire brindle bricks using 1.5:1 sand to Portland cement mortar. These are high-quality engineering bricks. The load-bearing walls separating the bascule chambers from the central stream of the river – which carry more than half the weight of the fixed steel girders carrying the roadways over the bascule chambers, and almost all of the weight of the bascules – are built entirely using Staffordshire brindle bricks. The piers are faced with granite blocks.

The masonry from 4ft above Trinity high water to road level was built by Perry & Co. under Tower Bridge Contract No. 7. It includes sixteen $2.44 \times 2.44 \times 0.91$m (8 × 8 × 3ft) blocks of granite on each pier, four each to support the spreading feet of each steel column. The lower surface of the granite blocks are cemented on to Staffordshire brindle brickwork 6.05m (19ft 10in) below street/footway level.

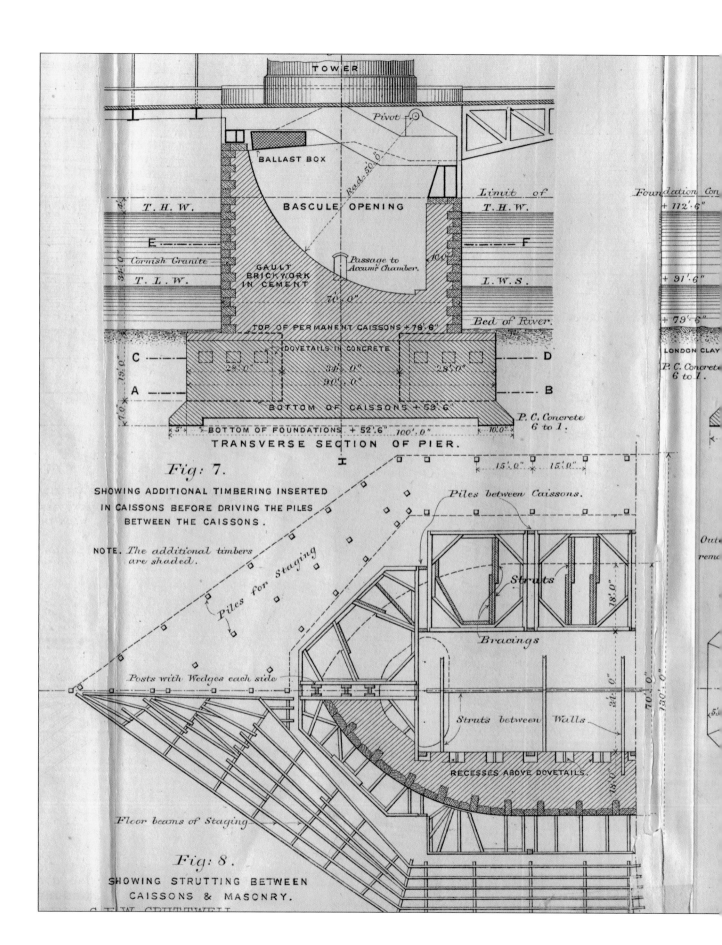

TOWER

Pivot

BALLAST BOX

Rad. 50'.0"

Limit   of

T.H.W.   T.H.W.

E   F

BASCULE   OPENING

Cornish Granite   10'0"

Passage to
Acum.r Chamber.

T.L.W.   L.W.S.

GAULT
BRICKWORK
IN CEMENT   70'.0"

Bed of River.

TOP OF PERMANENT CAISSONS +78'.6"

DOVETAILS IN CONCRETE

C   D
      28'.0"   30'.0"   28'.0"

A   90'.0"   B

BOTTOM OF CAISSONS +59'.6"

P. C. Concrete
6 to 1.

5'   BOTTOM OF FOUNDATIONS +52'.6" 100'.0"   10'.0"

TRANSVERSE SECTION OF PIER.

Fig: 7.

SHOWING ADDITIONAL TIMBERING INSERTED

IN CAISSONS BEFORE DRIVING THE PILES

BETWEEN THE CAISSONS.

NOTE. *The additional timbers
are shaded.*

15'.0"   15'.0"

Piles between Caissons.

Piles for Staging

Struts   18'.0"

Bracings

Posts with Wedges each side

Struts between Walls

Floor beams of Staging   RECESSES ABOVE DOVETAILS.

Fig: 8.

SHOWING STRUTTING BETWEEN

CAISSONS & MASONRY.

Foundation Con
+ 112'.6"

+ 91'.6"

+ 79'.6"

LONDON CLAY

P. C. Concrete
6 to 1

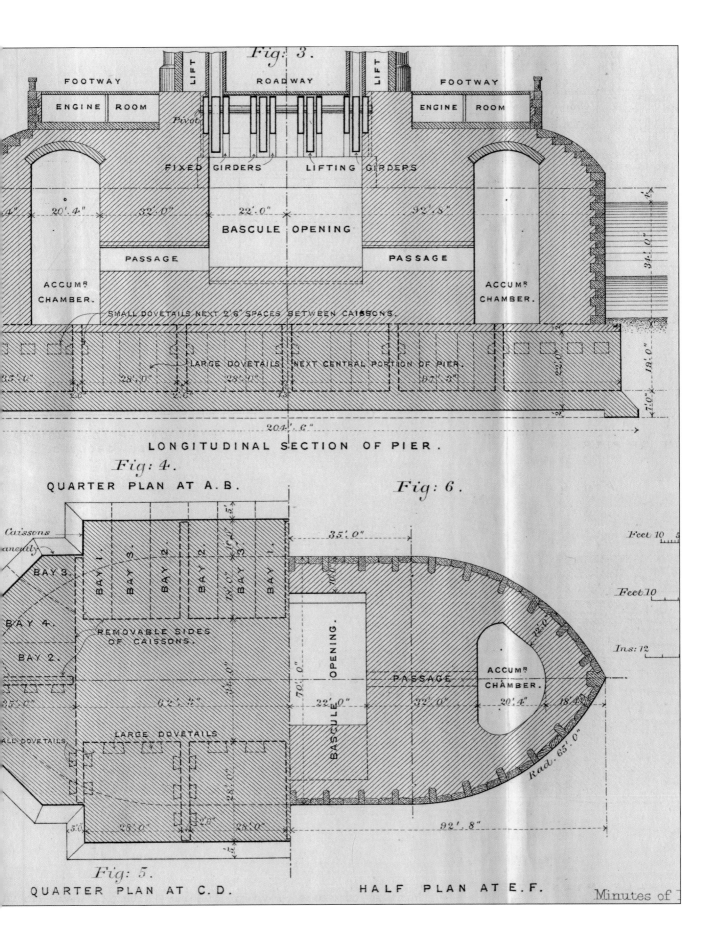

Fig: 3.

FOOTWAY   LIFT   ROADWAY   LIFT   FOOTWAY

ENGINE ROOM   ENGINE ROOM

Pivot

FIXED GIRDERS   LIFTING GIRDERS

20'.4"   32'.0"   22'.0"   92'.8"

BASCULE OPENING

34'.0"

PASSAGE   PASSAGE

ACCUMᴿ CHAMBER.   ACCUMᴿ CHAMBER.

SMALL DOVETAILS NEXT 2'6" SPACES BETWEEN CAISSONS.

LARGE DOVETAILS NEXT CENTRAL PORTION OF PIER.

19'.0"

35'.0"   28'.0"   28'.0"   28'.0"   97'.3"

204'.6"

LONGITUDINAL SECTION OF PIER.

Fig: 4.

QUARTER PLAN AT A.B.   Fig: 6.

Caissons placed obliquely

35'.0"

Feet 10

BAY 3.   BAY 1.   BAY 3.   BAY 2.   BAY 2.   BAY 3.   BAY 1.

BAY 4.

Feet 10

REMOVABLE SIDES OF CAISSONS.

BASCULE OPENING.

ACCUMᴿ CHAMBER.

Ins: 12

BAY 2.

70'.0"

PASSAGE

35'.0"   62'.3"   22'.0"   32'.0"   20'.4"   18'.4"

LARGE DOVETAILS

28'.0"

Rad. 65'.0"

SMALL DOVETAILS

35'.0"   28'.0"   28'.0"   92'.8"

Fig: 5.

QUARTER PLAN AT C.D.   HALF PLAN AT E.F.

The main towers

Each main tower features a riveted steel superstructure which supports not only the long chains and high-level ties of the suspension bridge, but also two cantilevered sections of high-level walkway and half the weight of each of the two central walkway sections suspended from the cantilevered walkways of the north and south main towers. The steel superstructure also supports some of the weight of the granite and Portland stone outer 'cladding' of the tower and all of the weight of the internal facing brickwork.

At each of the four corners of the structure is a riveted octagonal column 36.4m (119ft 6in) tall and 1.77m (5ft 9¾in) 'across flats'. Each column starts from a base, or 'foot', 4.3m (14ft) 'across flats'. This is designed to firmly plant and stiffen the column and transmit the weight it carries to its four granite foundation slabs below.

The bottom of the foot of each column is a flat octagonal steel base-plate, comprising three layers of ½in-thick steel plates riveted together, to which the webs, or gussets, of the base are riveted. The base-plate is bolted down to the granite foundation slabs and bedded using canvas and red lead. It is thus able to resist a degree of horizontal force applied to the columns between ground level and the top of the tower.

The centrelines of the columns are 18.4m (60ft 6in) apart transversely and 10m (33ft) apart longitudinally.

The columns are braced by three sets of four 1.83m (6ft) deep substantial riveted girders which directly support the weight of the internal brickwork and carry additional cross girders to support the internal landings and stairways of the tower.

The masonry below the first landing is mostly built directly on to the brick substructure of the pier.

Each column is constructed from long flat wrought-steel plates which form the straight sides of the octagon, and inner and outer joint plates, with an included angle of 135°, riveted to the side plates at the corners of the octagon. The plates vary in thickness between 19mm (¾in) and 22.2mm (⅞in). Between the base and the first landing, the side plates are tripled, between first and second landings doubled,

and between second and third landings single. The columns are stiffened internally by vertical T-section stiffeners riveted to the centre of the side plates, and by horizontal diaphragms positioned every 3ft. These have a central hole of either 762mm (2ft 6in) or 914mm (3ft) to provide access for inspection, maintenance and repainting, and to enable services to be routed up and down the columns.

At the top of each column is a heavy-duty diaphragm, consisting of a honeycomb of vertical plates, designed to transmit the applied load to the column. This is topped by multiple flat plates riveted together to bear the applied load.

A fourth landing, which supports the roof of the tower, is supported on steel stanchions, riveted to the tops of the columns. The stanchions are divided into two at their lower ends to allow the long chains, the outer girders of the high-level walkways and the high-level ties to pass between them. The stanchions are visible in the photograph at bottom right on page 59.

As the columns are now encased in masonry, the best way to understand the whole structure today is to study design drawings or photographs of the towers taken during construction. The photograph opposite shows the north tower in May 1892 just as Perry &

The main chains and the suspended roadway of the shore spans

Main chains

The main chains are braced steel girders, designed to have sufficient rigidity to prevent the suspended roadway from flexing under load. Each long chain has an 'eye' at both ends, one to form a pinned joint with the end of its high-level tie inside the neighbouring main tower, and one to form a joint with its short-chain neighbour. Each short chain is similar, one eye forming a pinned joint with its long-chain neighbour and the other connecting to its horizontal link inside the neighbouring abutment tower.

The long and short chains are linked above the suspended roadway by means of a pin 457.2mm (18in) in diameter, working in a split sleeve of outer diameter 685.8mm (27in).

The lower end of each long chain is formed into a solid eye 343mm (13½in) thick, working within a fork formed at the lower end of its short chain partner, the fork measuring approximately 686mm (27in) between its outer faces. The pin protrudes beyond the outer faces of the fork to carry a pair of thick forged plates which directly carry the weight of the transverse girder immediately below.

Co.'s men were starting to build up the masonry cladding. The lower parts of the near columns have been washed with neat Portland cement to prevent rust. They would be later wrapped with oiled canvas to prevent the masonry from adhering to the steel, thus allowing for expansion and contraction of the steel within its masonry jacket.

Each main tower included two hydraulic hoists (lifts) and two sets of stairs, each of 206 steps, to separate ascending and descending pedestrians.

RIGHT Looking up from the third landing (footway level) of the Surrey main tower. Cross girders are fitted between the main attic girders, and horizontal cross-bracing bars are fitted between each pair of cross girders. A cast-iron spiral staircase provides access from the third landing to the attic and roof, which is of all-steel construction. *(Author)*

The distance between the pin centres of a long chain (measured as a straight line, rather than along the curvature of the top boom) is 66.1m (217ft). The straight-line distance between the pin centres of a short chain is 33.8m (111ft). At its deepest point, a long chain is 4.7m (15ft 6in) deep, measured between the outer flanges of the top and bottom booms. For a short chain, this measurement is 3m (10ft).

The top and bottom booms of the chains are heavy-section riveted box girders with 762mm (2ft 6in) wide flanges. At horizontal intervals of 5.5m (18ft), top and bottom booms are linked by smaller section girders which are vertical when the chains are in position. Each of these vertical girders attaches to the top boom by means of a pair of forged steel plates, which pass through slots in the lower flange of the top boom and are riveted to the inside of the webs of the top boom. Each vertical girder attaches to the bottom boom in a similar manner, the forged steel plates protruding through slots in the bottom flange of the bottom boom to link, via a pin of diameter 152mm (6in), to a suspender rod which carries one end of a main transverse roadway girder. The forged steel plates also connect to cross-bracing girders.

Curiously, at the point where the cross

RIGHT The west Middlesex chains, pictured from the footway at the point of connection of the chains. *(Author)*

braces intersect, the drawing shows that there is a cast-iron casting; I think we can be sure that this casting is not under tension and is operating purely as a spacer. Another curious fact is that the upper booms of all eight chains on the bridge were filled with coke breeze concrete, to protect the steel from corrosion, but the voids in the lower booms were not. I rather imagine that this was due to the fact that gravity was against Arrol's men and they were unable to think of a way to insert the concrete into the space inside the lower boom.

I have estimated the weight of a long chain to be approximately 236 tons and that of a short chain approximately 134 tons.

The suspended roadway

Each long chain carries nine forged suspension rods, while each short chain carries five. These rods have a diameter of 140mm (5½in) or 152mm (6in), the larger-diameter rods being used where their length exceeds 7.3m (24ft). Each suspension rod is in two parts, thickened at the join and threaded, one rod with a right-hand thread and one with a left-hand thread. Cylindrical couplings, also left- and right-hand threaded, enable the effective length of each suspension rod to be adjusted.

The lower end of each suspension rod on the Middlesex bridge and of each rod supported by a Surrey long chain is pinned, just above roadway level and within the cast-iron parapet, by means of a pin of diameter 152mm (6in), to two 76mm (3in) thick plates, one each side. These plates extend downwards and are pinned by means of a further 6in-diameter pin to the central web of a transverse girder, which is thickened locally to the same thickness as the lower eye of the suspension rod.

The arrangement is different for rods supported by each short Surrey chain. These rods pass right through the centre of their stiffening girder and connect directly with their transverse girder via a pinned joint. The webs of these transverse girders are formed into two parallel sets of five riveted plates to receive the eye of the suspension rod.

There are 15 transverse girders, 14 carried by suspender rods and 1 carried below the pin joining the short and long chains. The girders are 19m (62ft 5in) long, 1m (3ft 3in)

deep at the centre, and weigh approximately 28 tons each. Longitudinal girders are fitted between each pair of transverse girders, spaced 2.3m (7ft 6in) apart. Transverse corrugated, pressed steel flooring 9.5mm (⅜in) thick and 152mm (6in) deep is riveted to the longitudinal girders. Rollers are provided where longitudinal girders meet the masonry of the piers and the abutments, and a number of expansion joints are provided.

The width between the parapets is 18.3m (60ft). As originally configured, the roadway was 11m (36ft) wide and there were two pathways, each 3.7m (12ft) wide. The roadway as originally completed comprised a 6in layer of coke breeze concrete (filling the troughs) plus an additional 3in of coke breeze concrete, supporting a bearing surface of wooden blocks. The pathways were constructed as 3in-thick stone paving supported on longitudinal coke breeze concrete walls, the spaces between the walls being used for gas, water, electrical and hydraulic pressure services.

Stiffening girders

In order to limit the tendency for loading on one shore span to cause vertical movement (hogging) in the other, possibly resulting in oscillation of the structure and excessive movement of the rockers in the main towers and abutment towers, stiffening girders were included in the design. These are fitted to the Surrey shore span only. The design objective was to limit the rise and fall at the pinned connections of Surrey long and short chains to 32mm (1¼in), this movement being due to the extension of the Surrey land ties and short chains due to the combination of traffic load and temperature changes.

A key reason for doing this was that high-pressure water supply pipes were installed under the east footway of the Surrey shore span and a flexible joint able to handle a 6in vertical movement would have been very difficult to implement. As a result of the stiffening girders, the vertical movement at the pinned connections of Middlesex long and short chains was increased and could amount to some 152mm (6in) when the bridge was opened. Seizing of pinned joints and rockers over the years is believed to prevent any hogging taking place today.

In principle, the west stiffening girder is a straight girder connecting the pinned joint of the west Surrey long and short chains to the box-girder at the base of the columns on the west side of the Surrey abutment tower; the east stiffening girder is the same.

In practice, and for purely aesthetic reasons, the structure is much more complicated. It was decided that the stiffening girders should be housed wholly within the parapets of the shore spans so as not to affect the 'lines' of the bridge. This complicated the design. Firstly, the stiffening girder cannot be attached directly to the centre of the pin linking the chains as it would have to rise above the parapet to do so. Secondly, the girder must allow suspension rods to pass through it. The design decided upon is shown below. It is clear that the effect of the design is that of a virtual

girder linking the pin connecting the short and long chains to the girder at the base of the abutment tower.

The success of the plan to 'hide' the stiffening girders can be judged by the fact that, when a civil engineering consultancy was commissioned to perform a detailed bridge assessment of Tower Bridge in 2012 as part of the Thames Tunnel project, the presence of the stiffening girders – major structural members of the bridge – went undetected!

The stiffening girders add substantial weight to the Surrey suspended span, increasing both the weight carried by the Surrey chains and the tension in the Surrey land ties.

BELOW Elevations and sections of stiffening girder. *(Tuit, ICE)*

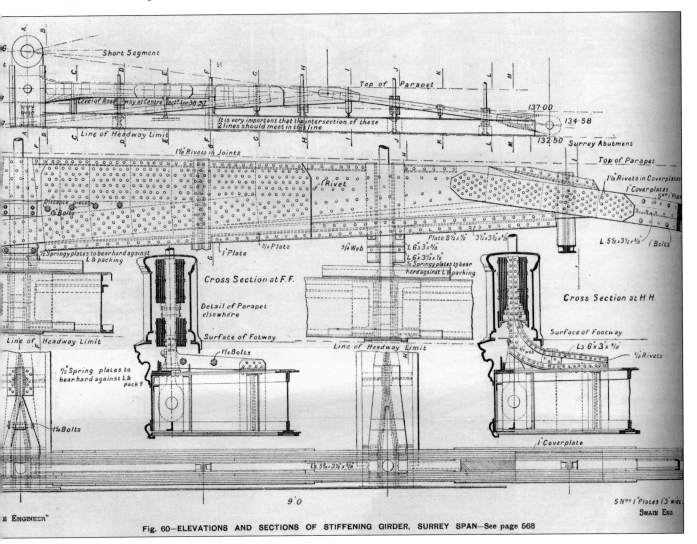

Fig. 60—ELEVATIONS AND SECTIONS OF STIFFENING GIRDER, SURREY SPAN—See page 568

The abutment towers

The foundation of each abutment tower is of mass concrete, 24m (80ft) long in the longitudinal direction of the bridge, approximately 41m (135ft) wide in the transverse direction of the bridge, and 4.5m (15ft) thick. The level at which the underside of the concrete meets the London clay is some 3.4m (11ft) below the level of the river bed in the centre of the river, 20m (66ft) below roadway level. Both abutment tower foundations were built inside coffer dams by John Jackson under Tower Bridge Contract No. 1. He also built the brick substructure

BELOW **The south side of the Surrey abutment tower.** *(Author)*

above the foundation slabs to a height of 4ft above Trinity high water. This is constructed predominantly of gault brickwork.

The Surrey abutment tower is wider than the Middlesex abutment tower and provides additional accommodation at ground floor level. The Surrey tower substructure also features a wide passage below and parallel to the roadway, through which coal, unloaded from barges moored against the abutment, could be transported to the coal store under the southern approach.

Each abutment tower contains a steel superstructure comprising four octagonal steel columns, 13.1m (43ft) tall and 1.45m

(4ft 9in) 'across flats'. The construction of these, including the stiffeners and diaphragms provided, is similar to that of the columns in the main towers. The centrelines of the columns in the longitudinal direction are 4.6m (15ft) apart. They share a common base, which takes the form of a very heavy longitudinal box girder of all-riveted construction with many strengthening webs. The girder is 7.6m (25ft long) 1.2m (4ft) high and 1.2m (4ft) wide. To provide transverse support to the columns, the box girder incorporates transverse box girder projections which are again 1.2m (4ft) high and 1.2m (4ft) wide. The transverse girders are approximately 2.8m (9ft 3in) long from end to end. Two

25.4mm (1in) thick octagonal plates are riveted to the top of the base at each column position to provide the platform upon which each column is raised.

The bases of the Surrey abutment columns are similar to the bases of the Middlesex abutment columns but they each include a pinned connection to their stiffening girder.

Each column base is bedded with Portland cement grout on to granite bed-stones and bolted to anchorages provided for the purpose.

Horizontal and diagonal (but not cross) bracing is provided between columns in the longitudinal direction of the bridge.

Transverse bracing between columns takes

the form of box girders, the lower flanges of which are formed in the shape of an arch.

The abutment steel columns have similar diaphragms to the main tower columns, including, at the top of each column, a diaphragm consisting of a lattice of vertical plates. Plates with a total thickness of 76mm (3in) provide bearing surfaces for the rockers.

The horizontal link comprises two thick steel plate assemblies spaced such that their inner faces are 610mm (24in) apart at the positions of the pins. These plates are 1.524m (60in) high by 6.7m (20ft) long, and are made from several 22mm (⅞in) thick plates riveted together, the thickness at the points where the pins are positioned being 156mm (6⅛in). The pin connecting the land tie to the link is 533.4mm (21in) in diameter. It is not provided with a split-sleeve bearing, or with lubrication, as little rotation was anticipated.

The river end of each link is connected to its short chain by means of another pinned connection, comprising a pin of 457.2mm (18in) diameter working in a split sleeve of outer diameter 685.8mm (27in). One half of the split sleeve is firmly clamped to the end of the short chain, thus preventing movement of the other half. The pin is drilled axially and also cross-drilled to provide a passage for oil, and oil grooves are machined into the working surface of the pin.

The outer masonry of the abutment towers is built directly on the abutment tower brick

substructure. Two-storey accommodation is provided above the arch, supported on wooden floors, the weight of which is borne by the masonry. The accommodation above the arch of the Surrey abutment tower was originally provided for the Bridge Master. It is used today as an educational space and has been used on occasion for wedding receptions. Superior and more convenient accommodation for the Bridge Master was provided in Edwardian times in the form of a three-storey house directly to the west of the southern approach, almost opposite the accumulator tower.

The Middlesex abutment tower includes a stone spiral staircase within a turret on its north-west corner. It is believed that the accommodation above the Middlesex abutment tower was intended for the Deputy Bridge Master, but there is some evidence that George Cruttwell occupied it for a time. As there is only one means of access, the space in the Middlesex abutment tower is used today for storage only.

The high-level ties

Each high-level tie is 91.7m (301ft) long and has a cross-section of 609.6mm (24in) by 203.2mm (8in) for most of its length, the width increasing to 267mm (10½in) at the positions of cover plates. It was made in two halves, each half consisting of four 25.4mm (1in) thick plates riveted together using countersunk head rivets on the inside faces. The two halves were then bolted together using 25.4mm (1in) bolts.

As originally designed and built, the dead weight of each tie (with the exception of a section at each end, including the eyes) was carried by 20 rods suspended from the top boom of the outer footbridge girder, these rods being spaced approximately 12ft apart. The weight of the section of high-level tie carried by each footbridge was approximately 78 tons – approximately the same as that of a large steam tank locomotive.

At each end of a high-level tie, the tie branches into an eye in the form of a fork. Each prong of the fork consists of riveted plates 1.7m (5ft 7½in) high and 203mm (8in) thick. The spacing between the prongs of the fork is 686mm (27in), the thickness of the upper eye of a long chain. The total weight of each high-level tie is approximately 120 tons.

The ends of the four long chains are pinned to the two high-level ties above the four landward columns of the main towers by means of pins 508mm (20in) in diameter, working in split sleeve bearings of outer diameter 762mm (30in). Lubrication of the pins in their plain bearings is provided.

The rocker assembly was designed with great care to ensure even distribution of the load transmitted by the arms of the 'fork' at the end of the high-level tie. It bears more than a passing resemblance to the landing gear of a jumbo jet.

A separate carrier beneath each arm of the fork splits the load transversely into two. These loads are then split again longitudinally, a quarter of the total load being carried by each of four thick plates. Each of these plates is fitted with two rockers 914mm (3ft) long, which transmit the load to the bearing plate. There is thus a total length of contact surface of 7.3m (24ft) over which in the region of 1.016×10^6kg (1,000 tons) of weight is transferred to the underlying column. The rockers have a working diameter of 508mm (20in) but have scalloped sides so that they can be spaced much more closely than their diameter would suggest, the distance between the centrelines of the outer rockers being just 1,016mm (40in).

The design ensures that all rockers remain parallel and cannot drift sideways. When the rockers had been installed, it was observed that extremes of hot and cold weather caused the rockers to move approximately 1in either side of their mean positions.

The land ties and anchor girders

All four land ties take the form of box girders above the level of the roadway so that they can support their own weight. Below ground, each Surrey land tie takes the form of a pair of horizontal flat ties of rectangular section, constructed from six 22.2mm (⅞in) thick plates riveted together. Each half of the tie is 838mm (2ft 9in) wide, according to Tuit, and 133mm (5¼in) thick, the two halves separated by 683mm (2ft 2⅞in).

ABOVE **The west Middlesex land tie entering the parapet.** *(Author)*

As each Surrey land tie approaches its anchor girder, it branches into three anchor ties, each of which is attached to its anchor girder by means of riveting.

Each Middlesex land tie, below ground, takes the form of a solid section constructed from 12 plates 533mm (21in) wide and 23.8mm (¹⁵⁄₁₆in) thick, disposed vertically. As

BELOW **The west Surrey land tie passing a support.** *(Author)*

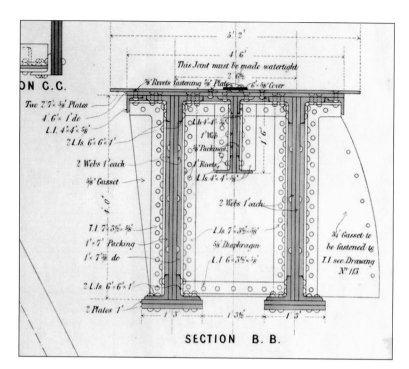

Labels on the left diagram:

ON C.C.

5'.2"

4'.6"

This Joint must be made watertight

2'.6"

⅝" Rivets fastening ⅝" Plate

6"×⅝" Cover

Two 2'.7"×⅝" Plates
4'.6"×1" do
L.I. 4"×4"×⅝"
2 L.Is. 6"×6"×1"
2 Webs 1" each
⅝" Gusset

L.Is 4"×4"×1" Web
⅝" Packings
1" Rivets
L.Is 4"×4"×⅝"

1'.6"

2 Webs 1" each

1'.0"

T.I. 7"×3½"×⅝"
1'×7" Packing
1'×7⅞" do

L.Is. 7"×3½"×⅝"
⅝" Diaphragm
L.I 6"×3½"×⅝"

¾" Gusset to be fastened to T.I. see Drawing No 115

2 L.Is. 6"×6"×1"
2 Plates 1"

1'.3"   1'.3½"   1'.3"

SECTION B. B.

LEFT This detail is taken from Contract Drawing No. 115 and shows a section through a Surrey anchor girder. The girder is a box girder, each vertical web comprising two 1in-thick plates. *(LMA)*

for the high-level ties, these ties were riveted in two halves, the halves then being bolted together with 25.4mm (1in) bolts. The resultant ties have a section of 533 × 286mm (21 × 11¼in), rising to 533 × 343mm (21 × 13½in) where cover plates are fitted. At its lower end, each Middlesex land tie is formed into an eye, to enable it to be pinned to an anchor tie by means of a forged steel pin of 584mm (23in) diameter. The anchor tie is riveted to its anchor girder. No precise details of these connections are available to the public, so a detailed description of the Middlesex anchorage cannot be given.

RIGHT Also taken from Contract Drawing No. 115, this image illustrates how an arm of an anchor tie is to be riveted to an anchor girder web. Two ⁵⁄₁₆in-thick packing sheets pad the web thickness out to the same 2⅝in thickness of the anchor tie arm. The anchor tie arm butts against the anchor girder web, two 1½in-thick steel plate links straddle both anchor tie arm and anchor girder web, and the whole assembly is riveted together with 1⅛in D rivets. *(LMA)*

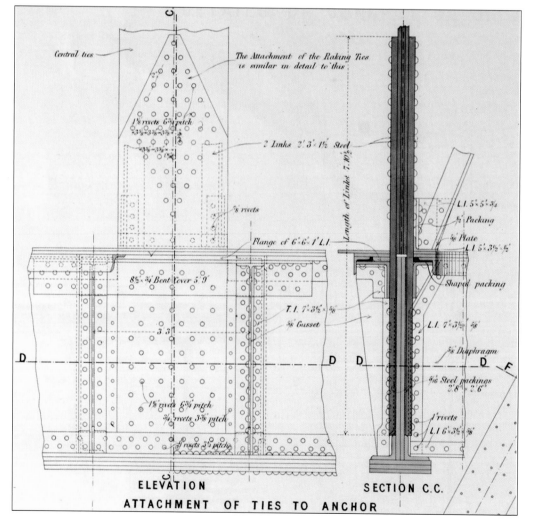

Labels on the right diagram:

Central ties

The Attachment of the Raking Ties is similar in detail to this

1" rivets 6¾" pitch
¾"×3¾×3¾"

2 Links 2'.3"×1½" Steel
Length of Links 7'.10½"

⅝" rivets

Flange of 6"×6"×1" L.I.

8½"×¾" Bent Cover 5'.9"

3'.3"

L.I. 5"×5"×¾"
½" Packing
⅝" Plate
L.I. 5"×3½"×⅝"

Shaped packing

T.I. 7"×3½"×⅝"
⅝" Gusset

⅝" Diaphragm

1" rivets 6¾" pitch
¾" rivets 3⅜" pitch

⁵⁄₁₆" Steel packings 2'.8"×2'.6"

1" rivets

1" rivets 3½" pitch

L.I. 6"×3½"×⅝"

D   **D D**   **D F**

ELEVATION   **SECTION C.C.**

ATTACHMENT   OF   TIES   TO   ANCHOR

Different subcontractors were used for the steelwork embedded under the approaches, John Jackson using Messrs Andrew Handyside & Co. of Britannia Ironworks, Derby, on the northern approach, and William Webster using Messrs Newton, Chambers & Co. of Sheffield on the southern approach.

Cruttwell states that the maximum tension in each Surrey land tie is 1,900 tons (1,930,489.3kg or 18,938.1kN) while that in each Middlesex land tie is 1,200 tons (1,219,256.4kg or 11,960.9kN), the difference being due to the effect of the stiffening girders. The Surrey anchor girders were therefore made a little larger than the Middlesex anchor girders. According to Cruttwell, the Middlesex girders were each 914mm (3ft) wide, 1.52m (5ft) deep and 11.6m (38ft) long.

The weight of the Surrey anchor block is estimated to be approximately 7,850 tons. This ignores the lightening caused by the various voids in the block, but it also excludes the weight of cast iron and steel buried in the block. The maximum pull of the two Surrey land ties is 3,800 tons, applied at an angle of 33.69°. The vertical component of the land tie pull is 3,800*sin (33.69°) tons, which equals 2,108 tons. The weight of the block is much greater than this, so the anchor block is well able to resist the vertical pull of the land ties – and this is before the weight of two whole bays of the masonry arch viaduct borne by the anchor block is taken into consideration.

The horizontal pull of the land ties is 3,800*cos(33.69°) tons, or 3,162 tons. The area of the side of the anchor block below the base of the brick pier on the river side of the anchor block is approximately 2,400ft^2, so the horizontal load on the London clay is 1.32 tons/ft^2, well within the capacity of the clay.

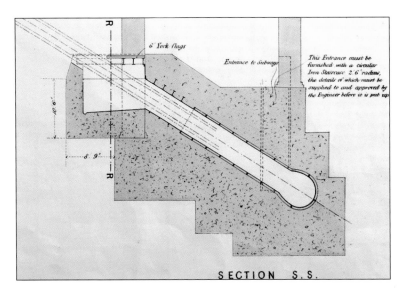

ABOVE This image is taken from Contract Drawing No. 113 and shows a section through the concrete anchor block on the centreline of the west central anchor tie. A cast-iron subway is cast into the anchor block to allow the anchor tie to pass through. The chamber at the top left of the image is also constructed from purpose-made cast-iron sections and houses the connection between a land tie and its three anchor ties. *(LMA)*

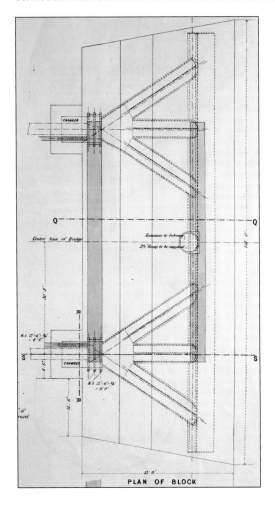

RIGHT A complete view of what is going on inside the anchor block. There are two chambers, six inclined anchor tie subways, an anchor subway running almost the entire length of the anchor block and a vertical subway, on the bridge centreline, fitted with a spiral staircase. There is a great deal of complexity, clever engineering design and cost hidden within the anchor block. Many patterns would have been required for the many differently shaped subway components. *(LMA)*

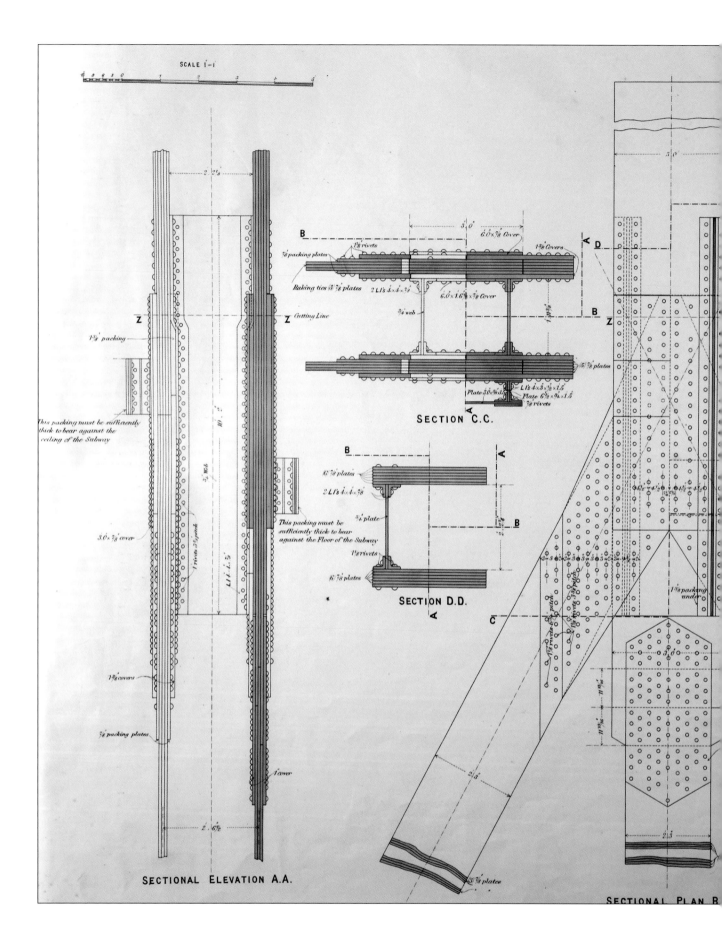

SCALE 1:1

This packing must be sufficiently thick to bear against the ceiling of the Subway

1⅜" packing

3.0" × ⅞" cover

1⅜" covers

⅞" packing plates

4" cover

SECTIONAL ELEVATION A.A.

B

⅞" packing plates

Raking ties (5) ⅞" plates

Cutting Line

2 L's 4" × 4" × ⅞"

1⅜" rivets

6.0" × ⅞" Cover

1⅝" Covers

6.0" × 1.6½ × ⅞" Cover

¾" web

A

D

B

Z

(5) ⅞" plates

Plate 3.0" × ⅝" × 1.5

L 1½ × 4 × 3 × ½ × 1.5

Plate 6½ × ¾ × 1.5

⅞" rivets

SECTION C.C.

B

6" ⅞" plates

2 L's 4" × 4" × ⅞"

¾" plate

A

This packing must be sufficiently thick to bear against the Floor of the Subway

1⅜" rivets

6" ⅞" plates

B

A

SECTION D.D.

1⅜" packing under

C

(5) ⅞" plates

SECTIONAL PLAN R

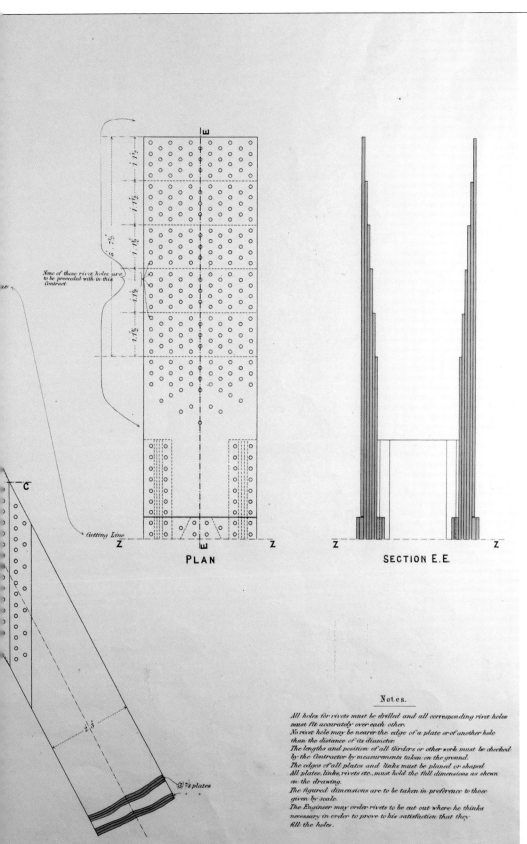

PLAN

SECTION E.E.

None of these rivet holes are
to be proceeded with in this
Contract

Cutting Line

C

(3) 7⁄8 plates

Notes.

*All holes for rivets must be drilled and all corresponding rivet holes
must fit accurately over each other.*
*No rivet hole may be nearer the edge of a plate or of another hole
than the distance of its diameter.*
*The lengths and position of all Girders or other work must be checked
by the Contractor by measurements taken on the ground.*
The edges of all plates and links must be planed or shaped.
*All plates, links, rivets etc., must hold the full dimensions as shown
on the drawing.*
*The figured dimensions are to be taken in preference to those
given by scale.*
*The Engineer may order rivets to be cut out where he thinks
necessary in order to prove to his satisfaction that they
fill the holes.*

LEFT Drawings for Contract No. 4 are available to the public and show that each Surrey anchor girder was 5ft 2in wide at the top, 4ft 3⅝in deep and 50ft long. Contract Drawing No. 116. This is a large drawing and reproducing it at a small scale means that much of the detail is lost. However, it shows the junction of a Surrey land tie with its three anchor ties. The incoming land tie comprises a pair of horizontally disposed flat ties as previously described. The only issue is that the width of the incoming land tie is shown as 3ft 0in, rather than the 2ft 9in given by Tuit.

We see two 'raking anchor ties' branching from the incoming land tie at an angle of 28.3°, plus a central anchor tie. The three anchor ties have the same section; they are each 2ft 3in wide and comprise an upper set of three ⅞in-thick steel plates and a similar lower set of three plates. The 'Sectional Elevation A.A.' in this drawing shows how this transition from the section of the incoming land tie is made. The 'Plan and Section E.E.' towards the right of the image shows how Newton, Chambers & Co. were required to finish the section of land tie falling within their subcontract – in such a way that William Arrol's men could form a connection with the length of land tie falling within Contract No. 6. None of the holes in the plan were to be drilled! *(LMA)*

93

THE SUSPENSION BRIDGES

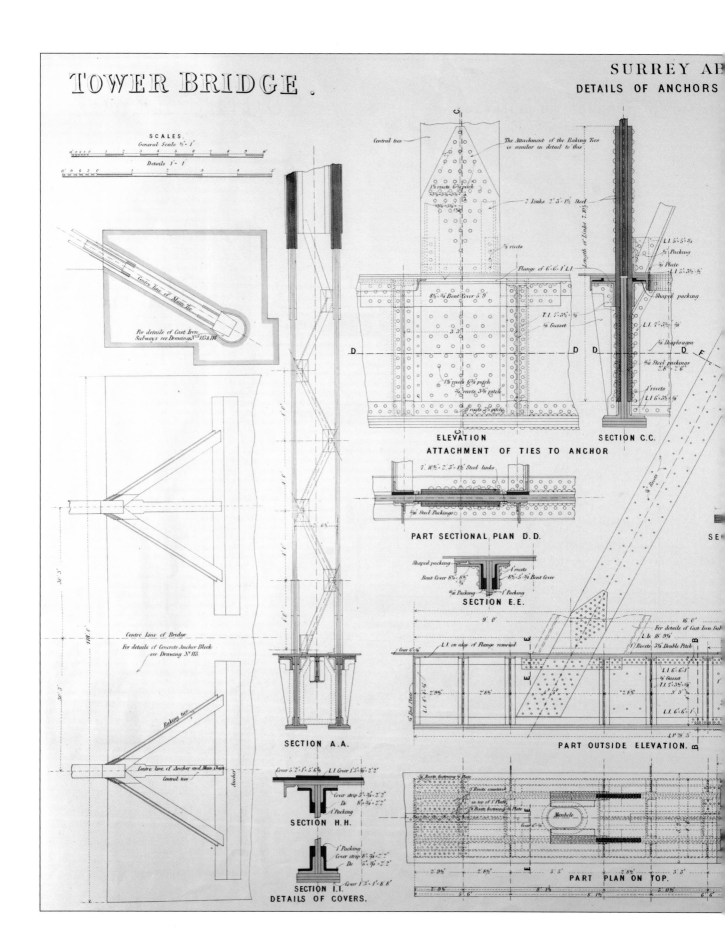

TOWER BRIDGE.

SCALES.
General Scale ½'- 1'

Details 1'- 1'

ELEVATION
ATTACHMENT OF TIES TO ANCHOR

SECTION C.C.

PART SECTIONAL PLAN D.D.

SECTION E.E.

SECTION A.A.

SECTION H.H.

SECTION I.I.
DETAILS OF COVERS.

PART OUTSIDE ELEVATION.

PART PLAN ON TOP.

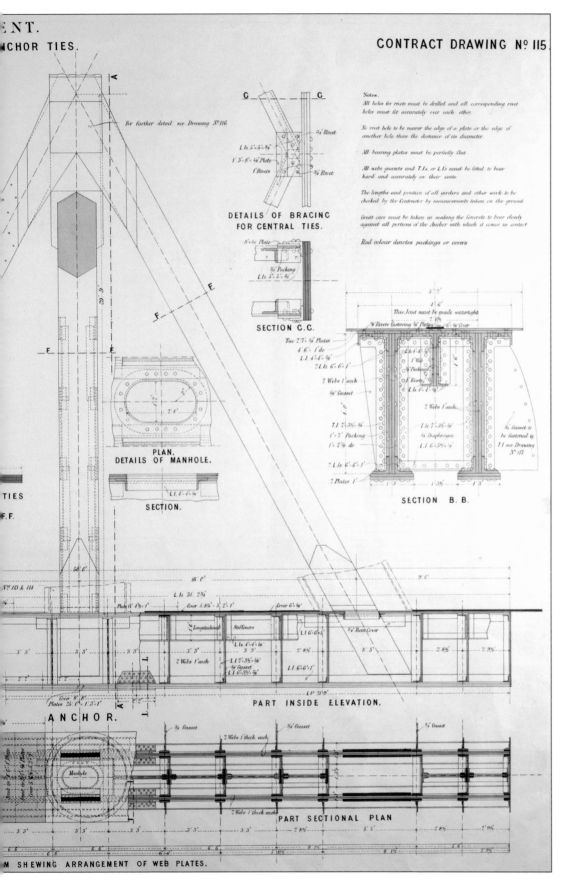

CONTRACT DRAWING № 115.

ENT.

NCHOR TIES.

For farther detail see Drawing № 116

C ────── C

¾" Rivet

L ls 5"x 5"x ¾"

1' 3"x 8"x ¾" Plate

1" Rivets

¾" Rivet

DETAILS OF BRACING
FOR CENTRAL TIES.

Notes.

All holes for rivets must be drilled and all corresponding rivet
holes must fit accurately over each other.

No rivet hole to be nearer the edge of a plate or the edge of
another hole than the distance of its diameter.

All bearing plates must be perfectly flat.

All webs gussets and T.ls, or L.ls must be fitted to bear
hard and accurately on their seats

The lengths and position of all girders and other work to be
checked by the Contractor by measurements taken on the ground

Great care must be taken in making the Concrete to bear closely
against all portions of the Anchor with which it comes in contact

Red colour denotes packings or covers

8"x ¾" Plate

¾" Packing
L.ls 5"x 5"x ¾"

SECTION C.C.

PLAN.
DETAILS OF MANHOLE.

L.I. 4"x 4"x ¾"

SECTION.

SECTION B.B.

50' 0"

16' 0"

9' 0"

Nos 113 & 114

PART INSIDE ELEVATION.

ANCHOR.

Manhole

PART SECTIONAL PLAN

M SHEWING ARRANGEMENT OF WEB PLATES.

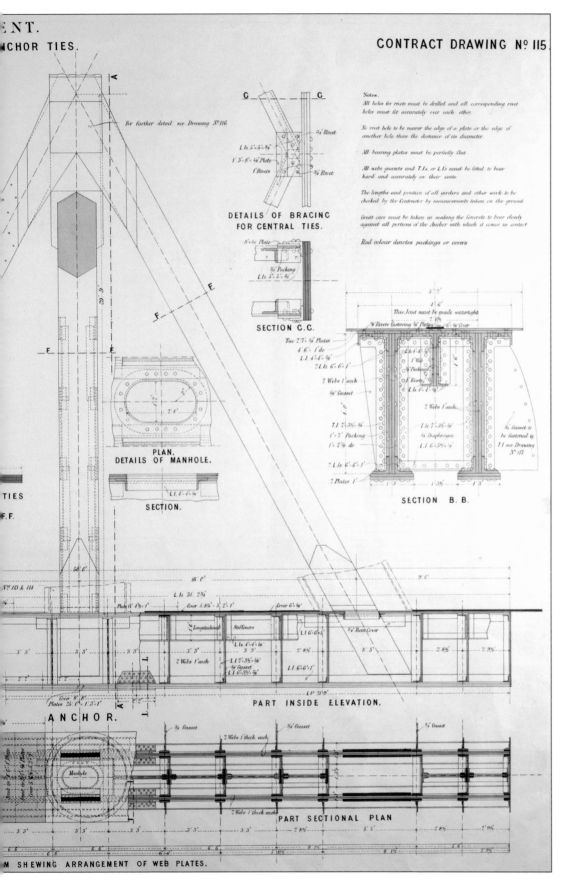

LEFT This is the whole
of Contract Drawing
No. 115. The total
cross-sectional area of
all twelve 1½in-thick
link plates attaching
three anchor ties to an
anchor girder is 486in^2.
The total pull of one
Surrey land tie is 1,900
tons, so the stress in
the link plates is 3.9
tons/in^2, well within
the tensile capacity of
the steel. *(LMA)*

TOWER BRIDGE

CONTRACT N⁰ 7

Middlesex Abutment Tower.

Scale 8 feet to an inch.

Half Elevation River Side.

Half Elevation Land Side.

Plan.

RIGHT AND OVERLEAF Two of George Stevenson's lovely drawings for Contract No. 7. They are signed by Perry & Co. *(LMA COL/CCS/ PL/01/002/A/313 and 304)*

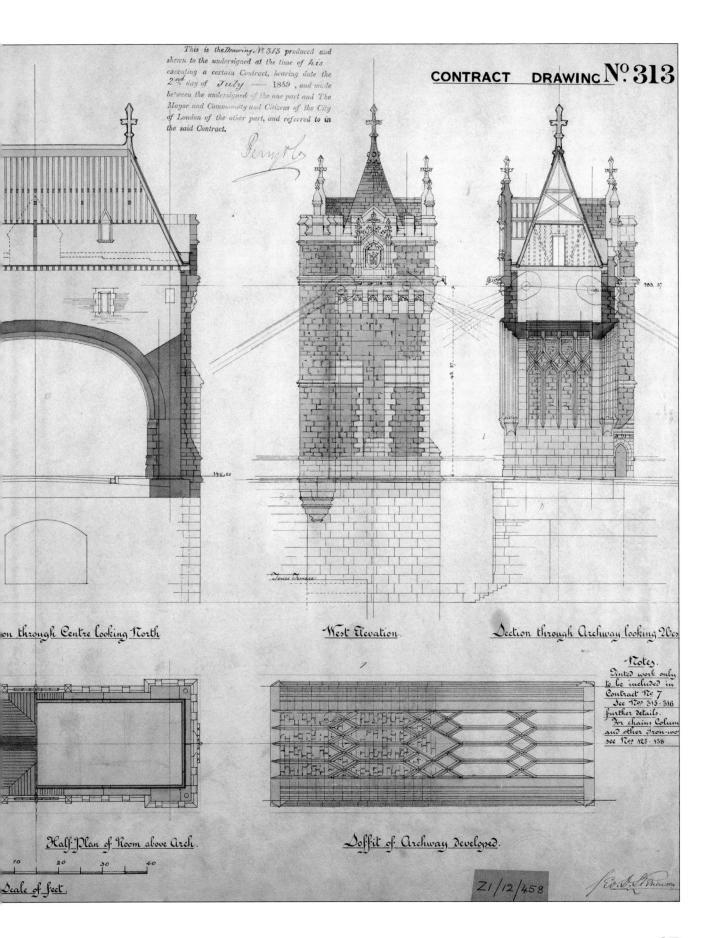

This is the Drawing No. 313 produced and shewn to the undersigned at the time of his executing a certain Contract, bearing date the 2nd day of July — 1859, and made between the undersigned of the one part and The Mayor and Commonalty and Citizens of the City of London of the other part, and referred to in the said Contract.

CONTRACT DRAWING No. 313

...on through Centre looking North

West Elevation.

Section through Archway looking West

Half Plan of Room above Arch.

Soffit of Archway developed.

Notes.
Tinted work only
to be included in
Contract No 7
See Nos 313. 316
further details.
For chains Columns
and other iron-work
see Nos 125. 158

10   20   30   40

Scale of feet.

Z1/12/458

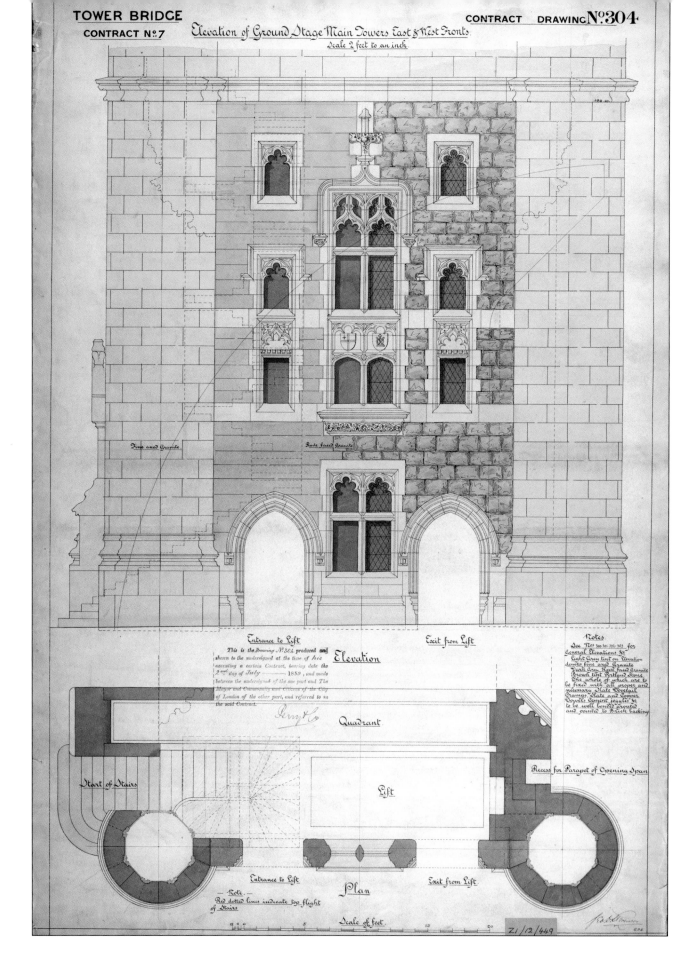

TOWER BRIDGE
CONTRACT Nº7

CONTRACT   DRAWING Nº 304

Elevation of Ground Stage Main Towers East & West Fronts.
Scale 2 feet to an inch.

The external masonry

All external masonry of both main towers and abutment towers above the level of Trinity high water plus 4ft was built by Perry & Co. under Contract No. 7. Materials used were grey Cornish granite, Portland stone and brick.

Ashlar (finely dressed) granite is used for the turrets of the main towers, for the horizontal bands marking the positions of the landings, for the arches of the main towers and abutment towers and for all quoins. Rock-faced granite is used for the walls. The granite was supplied by the De Lank quarry, St Breward, Bodmin. The stones were prepared and faced at the quarry using Brunton and Trier stone dressing and turning machines.

Portland stone is used for all windows, corbelling, balconies and parapets and for the ribs under the soffits of the abutment tower arches. Carving was subcontracted to Messrs Mabey and Son of Westminster.

Plain brickwork is used for all internal walls.

The forces in the structure

During the research undertaken for this book, many drawings of the structure as built were seen, but no information on the maximum load which the bridge was designed to carry came to light and no design calculations of any kind were seen.

The best we can do for the suspension bridges is to start from the three figures that *are* available. John Barry states in his 1894 lecture that the rockers under the pins of each high-level tie carry a load of 1,000 tons when the bridge is fully loaded. George Cruttwell, in his paper on the superstructure of Tower Bridge, gives this figure as 1,100 tons. This sounds a little less 'round', so we will use Cruttwell's figure. Cruttwell also gives the calculated maximum tension, *ie* when the bridge is fully loaded, in each Middlesex and Surrey land tie as 1,200 tons and 1,900 tons respectively.

I thought that it would be instructive and interesting to find out if I could produce my own estimates for these three parameters. As I don't have access to engineering analysis and design software, I thought I would deploy 'schoolboy/

schoolgirl statics' and available information on the elements of the structure. This exercise is described in Appendix 1. It is no substitute for a detailed computer-based stress analysis; it does, however, shed some light on the forces in the components.

After setting out the geometry of the bridge, the weight carried by each suspender rod on both the Middlesex and Surrey shore spans was calculated for four different scenarios:

1 The bridge as built in 1894, with no live load.
2 The bridge as built in 1894 with the test load of 1cwt per square foot across the entire roadway and footway area of a shore span.
3 The bridge in 2019 under no load (the bridge having put on additional weight since 1894, following several resurfacing projects).
4 The bridge in 2019 under an assumed maximum load.

The key conclusions are that:

1 The test load of 1cwt per square foot of shore span surface (810 tons) applied to a shore span in 1894 *was* the maximum design load.
2 The analysis yields figures for the maximum weight on a main tower column and the maximum tension in a Middlesex land tie which agree very closely with the Cruttwell figures.
3 The figure for the maximum tension in a Surrey land tie is very much less than the Cruttwell figure of 1,900 tons, indicating that the Surrey land ties were pre-tensioned so as to give the tension of 1,900 tons when all staging below the main chains was removed.
4 The bridge in 2019, even under maximum permitted live load, is still operating well within the original maximum load design envelope.
5 Under 2019 maximum load, defined as twelve 18-tonne lorries on a shore span:
    ◆ The maximum tension in each Middlesex horizontal link and in each high-level tie is 988 tons.
    ◆ The maximum tension in each Middlesex land tie is 1,188 tons.
    ◆ The maximum load transmitted to each landward column by the rockers at both ends of each high-level tie is 1,024 tons.

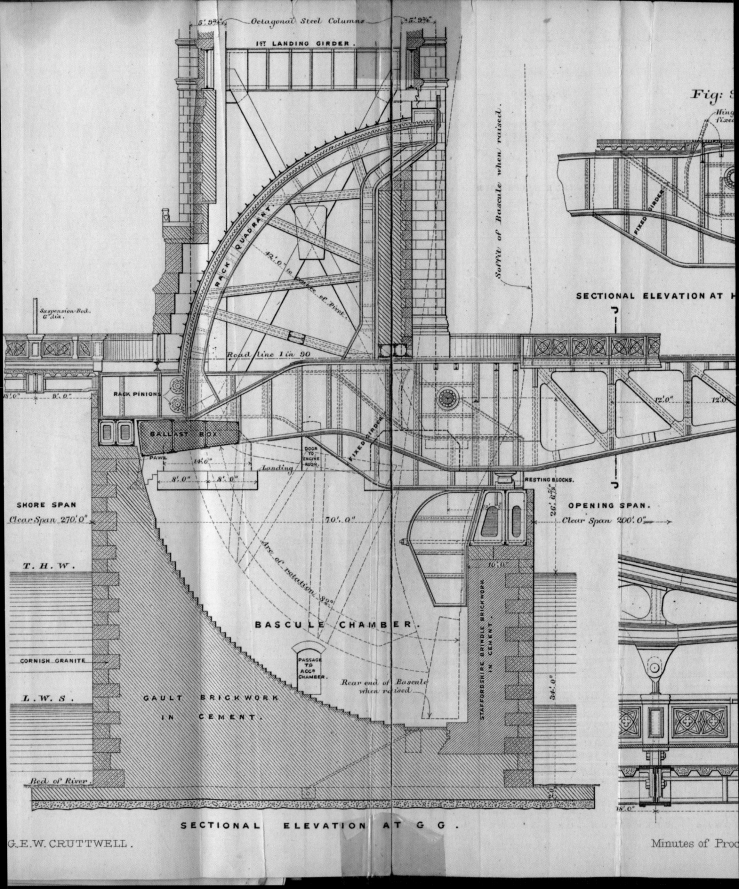

Octagonal Steel Columns

5' 9¾"   5' 9¾"

1ST LANDING GIRDER.

Fig: 9

Wing
fixed

Soffit of Bascule when raised.

SECTIONAL ELEVATION AT H

RACK QUADRANT

42'.0" to centre of Pivot.

FIXED GIRDER

Road line 1 in 90

RACK PINIONS

Suspension-Rod.
6" dia.

18'.0"   9'.0"

BALLAST BOX

PAWL

14'.0"

8'.0"   8'.0"

Landing

DOOR
TO
ENGINE
ROOM.

FIXED GIRDER

RESTING BLOCKS.

12'.0"   12'.0"

26'.6⅞"

SHORE SPAN

Clear Span 270'.0"

70'.0"

OPENING SPAN.

Clear Span 200'.0"

T.H.W.

Arc of rotation 82°

10'.0"

BASCULE CHAMBER.

CORNISH GRANITE

PASSAGE
TO
ACC.
CHAMBER.

STAFFORDSHIRE GRINDLE BRICKWORK IN CEMENT.

36'.0"

L.W.S.

GAULT BRICKWORK
IN   CEMENT.

Rear end of Bascule
when raised

Bed of River.

18'.0"

SECTIONAL   ELEVATION   AT G.G .

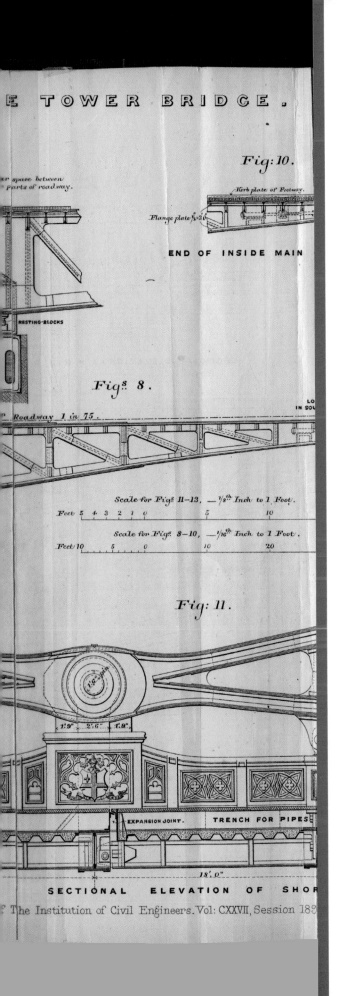

E TOWER BRIDGE.

Fig: 10.

r space between
parts of roadway.

Kerb plate of Footway.

Flange plate ⅝×3 0

END OF INSIDE MAIN

RESTING-BLOCKS

Figs 8.

Roadway 1 in 75.

Scale for Figs 11–13, — ⅛th Inch to 1 Foot.

Feet 5 4 3 2 1 0   5   10

Scale for Figs 8–10, — 1/16th Inch to 1 Foot.

Feet 10   5   0   10   20

Fig: 11.

EXPANSION JOINT.   TRENCH FOR PIPES

18' 0"

SECTIONAL   ELEVATION   OF   SHOR

Chapter Five

The bascule bridges

Each bascule is 162ft 3in long and weighs approximately 1,100 tons. Almost perfectly balanced, it rotates in roller bearings on a 21in diameter steel shaft. When fully open, a 200ft channel is available to shipping. The bridge has operated nearly 500,000 times. By law the bridge had to be raised on demand, 24/7. Nowadays, bridge lifts are requested by email and 24 hours' notice is required.

OPPOSITE **This is the sectional elevation of a bridge pier appended to the paper George Cruttwell presented to the Institution of Civil Engineers of 10 November 1896.** *(ICE)*

One bridge or two?

This is not a question of great significance as, even if we consider Tower Bridge to have two bascule bridges, they always operate as one. However, it is worth noting that a great many bascule bridges have only one bascule. Single bascule bridges were, and remain, popular for spanning dock entrances and canals, rather more 'industrial' locations than that of Tower Bridge. They tend to also have a very industrial look and feel, with much ironwork, including the counterweight, on display. A single-bascule bridge offering a 200ft channel for shipping would be huge and extremely unsightly and we must be grateful that one was never proposed for, or built at, the Tower Bridge site.

The bascule principle

Bascule is a French word meaning see-saw. The principle is a very simple one. A perfectly balanced bascule, though it might weigh almost 1,100 tons, does not require a massive amount of motive power to operate, once the initial inertia has been overcome. Of course, wind pressure on the bascules is a major factor, and the original machinery for Tower Bridge was designed to operate the bridge even in an extreme wind pressure of 56 pounds per square foot.

The second largest Meccano set in the world

The largest had been that for the Forth Bridge, but this one was still enormous. Perhaps Frank Hornby took his inspiration from it, Meccano being invented in 1898, just four years after the bridge was completed.

Contract No. 6 covered the provision, shipment, erection and painting of all the steelwork and ironwork of the bridge, with the exception of the anchor ties and the four anchor girders, provision for which had been made in Contracts 2 and 4 for the northern and southern approaches respectively.

The Schedule of Quantities and Prices in Contract No. 6 lists, by component, the weight of wrought and cast steel, wrought and cast iron, steel forgings and lead required for the building of the superstructure of the bridge, including the abutment towers. It amounts to 12,291 tons of iron and steel and 580 tons of lead. There was also an allowance for 550yd^3 of concrete to fill box girders, the main chain booms and the land ties. This was seen as an important measure to prevent the corrosion of inaccessible parts of the structure. The steelwork associated with the girders supporting the roadway over the bascule chambers and providing support for the main pivot shaft of the moving leaf amounted to 410 tons per pier plus 6.5 tons per pier of bolts and screws.

The steelwork was shipped from Sir William Arrol & Co.'s Dalmarnock Iron Works in Glasgow to a point on the Thames about half a mile downstream of the site of the bridge, using steamers of the Clyde Shipping Company and the Carron Company. From there, the steel was conveyed in barges to the site. Generally, between 50 and 100 tons of steel was delivered to London each week. Each piece of steel was stamped to identify its position in the bridge superstructure. Examples of these numbers can be seen today on the booms of the main chains.

Some fabrication was done in Glasgow but

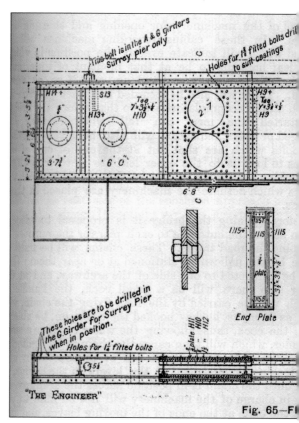

Fig. 65—FI

the weight of each sub-assembly to be shipped was limited, by choice, to about 5 tons. This meant that even girders of relatively modest proportions, such as the c25-ton transverse girders of the suspended shore spans, were assembled and riveted on site, Arrol's men working in the open air on the staging, on the approaches or on the banks of the river in all weathers. It is not clear why this decision was made, as it would have been far more convenient and pleasant for fabrication to have been undertaken under cover in the sheds of the works in Glasgow, with ready access to heavy machinery and cranes.

The steel girder bridges spanning the bascule chambers

A simply supported steel girder bridge carries the roadway over each bascule chamber. Granite bed-stones were bedded directly on to the Staffordshire brindle brickwork laid by John Jackson's men on the river side of each pier. Further brickwork was built on to the shore side of

each pier by Perry & Co., allowing similar granite bed-stones to be installed on that side too.

Holding-down bolts on the shore side of each pier were put in place and connected to the anchorages installed under Contract No. 1. Two box girders about 50ft long were then erected immediately above the granite bed on the shore side of each pier. They were then lowered on to the bed-stones, which had been covered with three layers of canvas thoroughly coated with red lead. More red lead was forced under the lower flanges of these box girders to ensure that the entire area of each bottom flange was providing a load-bearing surface. The holding-down bolts were then tensioned to 5 tons/in^2 and the nuts tightened – the Victorian equivalent of using a very large torque wrench.

The c50ft-long box girders atop the river side of each pier are much larger than the shore-side girders as they transmit almost all of the load of the moving leaf to the river wall of the pier. They were also fabricated above their bed-stones and lowered to within 1in of their final position. Portland cement grout was then injected to provide a bearing surface equal to the total area of the bottom flanges of the girders.

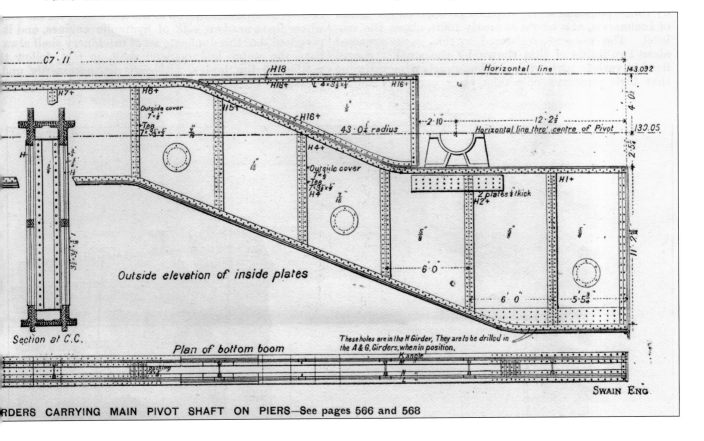

RDERS CARRYING MAIN PIVOT SHAFT ON PIERS—See pages 566 and 568

The fixed girders

There are eight fixed girders in each pier. They are simply supported, but their design is not simple, as the upper flange of the river end of each girder must be below, and provide support for, the pivot for the moving leaf, while the lower flange of the shore end of six of the eight girders must be above the ballast box of the moving leaf. This necessitates a joggled design. The sectional elevation drawing on the previous page shows the profile of the two outer fixed girders.

The two shafts which drive the pinions engaged with the racks attached to each of the outer main girders of the moving leaf pass through the fixed girders, bearings being provided on the two outer and the two centre fixed girders.

The fixed girders for each pier were built in pairs on the staging below the associated shore span and then drawn forwards and riveted into position. A vertical plate is riveted to the end of each fixed girder at the shore side of each pier. At the river side of the pier, each pair of fixed girders with no moving leaf between them is also riveted to a vertical plate. 'Floor' plates are installed between the lower flanges of each pair of fixed girders *not* having a moving girder between them. This provides a means by which the fixed girders can be inspected and maintained, manholes being provided at regular intervals in the girders to allow the repainting of internal surfaces. The manholes are very small; clearly this was a job for a boy at 4d per hour!

The cantilevers providing support under the main bearings of the moving leaf

The six cantilevers in each pier serve two purposes. Firstly, they provide additional support under the main bearings of the moving leaf, the centreline of the pivot shaft being 3ft 3in inside the inner wall of the river side of the pier. Secondly, they provide protection against the wall of the pier being damaged by a ship strike.

Each cantilever is riveted to the girder atop the river wall of each pier and also attached to it by means of a large bolt. Adjacent pairs of cantilevers are also riveted to, and braced by, two horizontal box girders.

The triangular box girders

Triangular-shaped box girders are riveted to the sloping upper flanges of the six inner fixed girders to support the roadway. Other

BELOW Cantilevers in the Surrey bascule chamber. The lower flanges of the six inner fixed girders and the floor plates between them are clearly visible. *(Author)*

OPPOSITE This photograph shows the Middlesex main tower of the bridge under construction. Perry & Co. have installed masonry up to a point halfway between road level and the first landing of the tower. The outer box girder supporting the fixed girders at the river side of the pier can be seen, as can the ends of the bolts attaching the two girders to the cantilevers. The eight fixed girders are visible, as are the triangular box girders carrying the roadway and the triangular box girders to which the bearing housings are bolted. The rear ends of the two outer main girders of the moving leaf have been fabricated on the staging adjacent to the pier (weighing between 50 and 60 tons each) and moved forward into position. A temporary mandrel has been installed through each moving girder and its adjacent bearings and the rear end of the girder gently rotated down into the bascule chamber. The two centre main girders of the moving leaf are being fabricated on the staging. We have a good view of the triangular girders carrying the steel arch of the main tower. We even catch a glimpse of the pivot shaft protruding from the pivot tunnel at the bottom right of the photograph. Keys are visible on the shaft. *(LMA/Collage 32 3342)*

A SIMPLE FORCE DIAGRAM OF A BASCULE

A simple force diagram of a bascule
(Both images, Author)

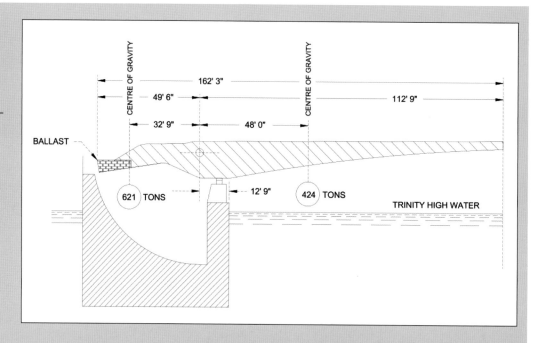

Above is a modern copy of a diagram included in John Wolfe Barry's 1894 lecture on the bridge. He wrote his lecture in 1893, before the bridge had been completed and tested. He includes a 'health warning' under the diagram as follows: 'The various weights given above are matters of estimate. They may be slightly modified in execution.' However, they give us a statement of intent with regard to the level of balance intended to be achieved.

Below is a simple force diagram for a bascule. For the bascule to be in perfect balance, the reaction (R2) at the resting blocks would be zero, the anti-clockwise moment around the pivot caused by the rear of the bascule and the ballast box, perfectly matching the clockwise moment of the opening leaf. Taking moments about the pivot:

621*32.75 + R2*8.75 = 424*48, from which we find that R2 is 1.63 tons. Clearly, this makes sense. The bascule must be slightly out of balance in order for it, when closed, to sit down firmly on the resting blocks to prevent traffic passing over the bascule from causing it to lift and bounce.

Horizontal steel girders attached between the fixed girders either side of the moving girders provide additional 'stops' for the ballast box of each bascule.

It is interesting to note that Barry tells us that the centreline of the pivot is 12ft 9in inside the outer wall of the pier, while George Cruttwell informs us (see the drawing on page 100–1) that this dimension is 13ft 3in. (The official Tower Bridge contract drawings support Cruttwell.) Also, Barry tells us that the total weight of ballast in each ballast box is 422 tons, while Cruttwell gives the weight as 365 tons. Clearly, achieving the right level of balance was a matter of trial and error. What we can be sure of is that successive Corporation of London engineers with responsibility for the bridge have endeavoured to prevent the moving leaves from gaining weight when they are resurfaced.

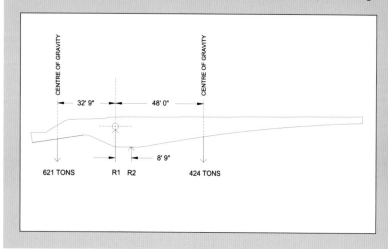

triangular-shaped box girders are attached to the horizontal top flanges at the river end of the six inner fixed girders. The six inner main pivot bearing housings are bolted to these.

The roadway

Lindsay's patent trough floor plates, mounted transversely across the inner six fixed girders and the upper flanges of the triangular box girders, support the roadway over the bascule chamber. The troughs were filled with concrete and a further 6in of concrete was laid on top to support the wood block roadway surface. The two middle main girders of the moving leaf were sculpted to sit below the lower surface of the trough floor plates.

Recent changes to the roadway construction over the bascule chambers

In 1993, the roadway over the bascule chambers was taken up, the underlying steelwork repaired and repainted and the roadway replaced. Reinforced concrete was laid down. It is unclear whether replacement trough flooring was first installed, or whether the concrete was laid directly on top of the fixed girders.

Bascule design

The bascules were designed such that, when closed, they give the opening part of the bridge the appearance of a shallow pointed arch. When open, each moving leaf sits inside the vertical line of the side of its pier, making the full channel width of 200ft available, with a clear headway of 140ft – 5ft higher than the headway prescribed in the Tower Bridge Act. Each bascule moves through an arc of 82°.

The elevation drawing from Cruttwell's paper on the superstructure of the bridge, shows that there are four main girders in each opening leaf. These are spaced 13ft 6in apart, enabling a clear width between the parapets of 49ft by bracketing out from the two outside main girders. As originally configured, there was a 32ft-wide roadway and two footways of 8ft 6in each.

The main girders take the form of box girders within the bascule chambers, each sitting between a pair of the fixed girders spanning the

BELOW The opening leaves. *(Author)*

ABOVE The underside of a moving leaf. *(Author)*

RIGHT Half transverse section of a moving leaf. *(Cruttwell, ICE)*

BELOW This photograph was taken from the bottom of the Surrey bascule chamber, showing the rear of the Surrey moving leaf. The ballast box, sitting below the fixed girders, and the two pawls supporting the rear of the leaf are clearly visible. *(Author)*

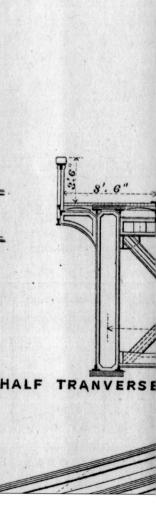

HALF TRANVERSE

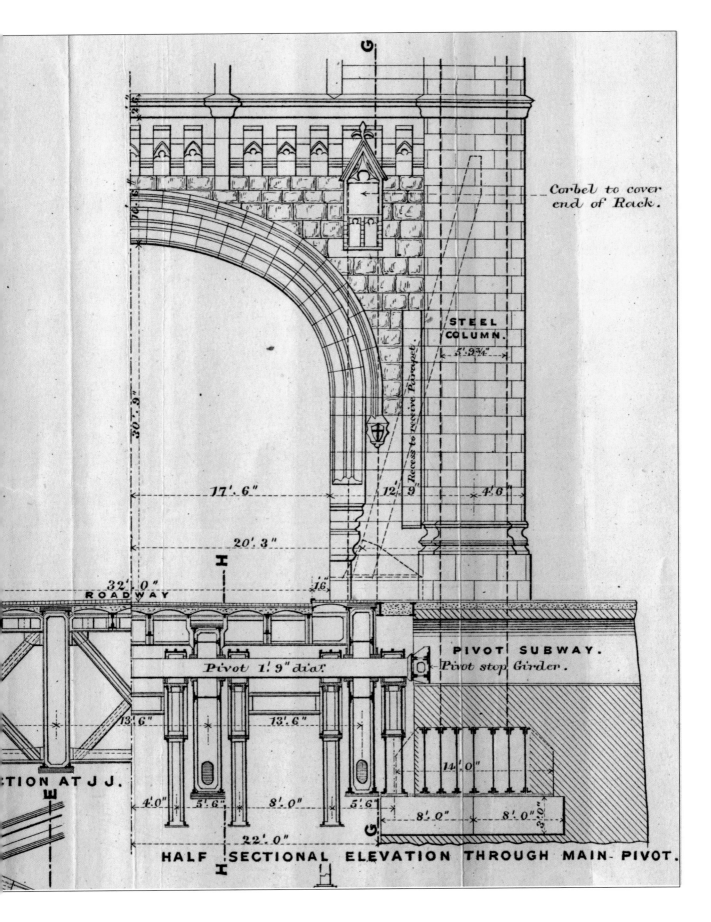

Corbel to cover
end of Rack.

STEEL
COLUMN.
5'.9¾"

Recess to receive Parapet.

17'. 6"

12'. 9"

4'. 6"

20'. 3"

32' 0"
ROADWAY

16"

30' 9"

74' 8"

2' 6"

PIVOT SUBWAY.

Pivot stop Girder.

Pivot 1' 9" dia.

13' 6"

13' 6"

14'. 0"

TION AT J J.

4'.0"

5'.6"

8'.0"

5'.6"

8'.0"

8'.0"

3'.0"

22'. 0"

HALF SECTIONAL ELEVATION THROUGH MAIN PIVOT.

bascule chamber, and of open lattice girders over the central stream of the river.

The lattice girders feature verticals every 12ft, with diagonal ties between them. Transverse girders and cross bracing are fitted between the main girders at the same spacing. Longitudinal girders are mounted on the upper transverse girders, and small transverse girders are fitted between these.

Quadrants and racks

The outer two main girders of each moving leaf carry steel quadrants, fabricated in sections at the Dalmarnock Iron Works and completed on site, to which cast steel rack segments are bolted. The individual rack segments are 17in wide and about 6ft long. There are two racks on each quadrant, the pitch of the teeth being 5.9in. Fixing the quadrants in place proved very problematic, owing to the confined space around the quadrants. Great care was taken to ensure that the quadrants were pure arcs of a circle (of radius such as to provide a radius of 42ft at the pitch line of the racks and pinions), centred on the pivot centreline. This required repetitive rotation of each partly built moving leaf to achieve a consistent spacing between the outer flanges of the quadrants and the pinion shafts. For this purpose, 37 tons of temporary lead counterbalancing was placed at the incomplete river end of each moving leaf and steam winches were installed to rotate the partly built moving leaves when this was required.

The steel rack segments were supplied by Sir W.G. Armstrong, Mitchell & Co. Ltd under Contract No. 5. To fit the rack segments, which weigh some 14 tons 8cwt per moving leaf, lines were scribed on the outer flange of the quadrant as it was rotated, and shims installed under the rack segments as needed, using the scribed lines as a reference.

Each quadrant works within a steel enclosure designed to protect the racks and pinions against the ingress of dirt and to protect bridge personnel from danger.

The sectional elevational drawing on page 100–1 shows that space within the main towers for the quadrants was, and remains, extremely limited. The masonry of the main tower facing

the shore spans had to be built out on both sides of the roadway to provide room for the quadrants, and the top of the quadrants, when the moving leaves are in the closed position, protrude through the river-facing masonry of the main towers and are protected by decorative castings.

Pinion shafts, pinion shaft bearings and pinions

This machinery was all supplied under Contract No. 5. There are two pinion shafts, one above the other, in each pier. The upper shaft in the Middlesex pier is driven from the east engine room and the lower shaft from the west engine room. The upper shaft in the Surrey pier is driven from the west engine room and the lower shaft from the east engine room.

Each shaft is in two parts with a coupling situated between the two central fixed girders spanning the bascule chamber. Live roller bearings are installed in the two outermost and the two central fixed girders. The holes in the two outermost fixed girders are large enough to enable a half-shaft, with its two 13-tooth pinions attached, to be withdrawn into the adjacent engine room. On the way to its engine room, each pinion shaft passes close by a main tower column. Just inside its driving engine room, each pinion shaft is supported by a bearing held by a floor-mounted bearing standard. A 29-tooth pinion is attached to the end of the pivot shaft, engaging with a pinion on an intermediate shaft. The reduction gearing between the crankshafts of the hydraulic engines and the pivot shaft is approximately 6.1:1.

Pivot shafts and main bearings

These were provided by Sir W.G. Armstrong, Mitchell & Co. Ltd under Contract No. 5. Each steel pivot shaft is 21in in diameter and 48ft long, and weighs some 25 tons. Each shaft is keyed to its four main moving girders by means of cast steel flanges bolted to the webs of the girders.

Each of the eight main bearings features a split, cast steel, housing. The bearing housings for the outermost fixed girders are bolted directly to the upper flanges of the girders. The remaining bearing housings are bolted to the triangular box girders attached to the six inner fixed girders, necessitating a different design. Live steel rollers, 22½in long and 4.4375in in diameter, work within the bearing housings.

Roadway construction

As originally constructed, ⅜in-thick steel floor plates were mounted longitudinally between the top flanges of the bascule main girders and bolted to the longitudinal girders at the top of each bascule. The upper surface of the floor plates was cylindrical, with a 3in camber, providing additional strength at no additional weight.

Shaped lengths of creosoted Memel pine (slow-grown *pinus sylvestris* shipped from the Baltic port of Memel), were laid transversely across the shaped floor plates, their ends butting up to the upper flanges of the main girders.

Two-inch-thick planks of greenheart (*chlorocardium rodiei* – a South American hardwood) were laid longitudinally on to the Memel pine and main girder top flanges, bolts securing the Memel pine and greenheart to the floor plates. (It is not clear how the greenheart planks were secured to the steel flanges of the main girders.) Greenheart is extremely durable and well suited to marine environments. It is also heavy, difficult to work and can cause wheezing, cardiac and intestinal disorders, severe throat irritation, and the tendency for wood splinters to become infected – hazards to be added to the long list of dangers already facing the men building the bridge.

Patent ACME paving blocks of creosoted Memel pine were laid diagonally on top of the greenheart planking, the blocks secured to each other by oak dowels incorporating spacers and to the greenheart by coach screws.

This joinery work was all conducted with the moving leaves in a vertical position – not an easy task. With the bascules lowered, asphalt was run into the ¼in-wide joints between the paving blocks.

The footways were constructed a little differently. The paving blocks were thinner than those used on the roadway, and they were

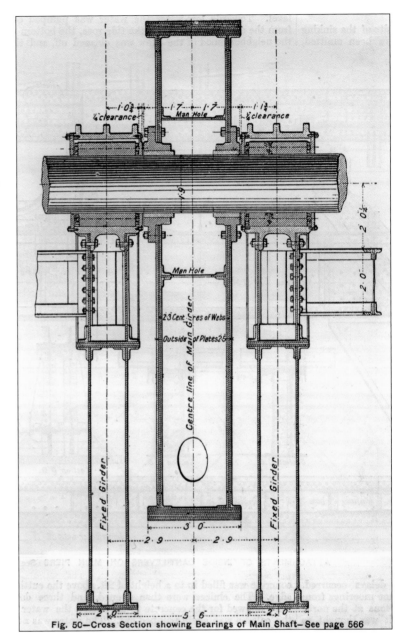

Fig. 50—Cross Section showing Bearings of Main Shaft— See page 566

ABOVE Cross-section of a pivot shaft showing the bearings – eight per pier. *(ICE)*

laid directly on to the Memel pine planks, the greenheart being omitted.

The roadway and footway on the moving leaves has been replaced several times since the bridge was built, water ingress being a constant problem, causing any wood employed in the construction to swell and gain weight.

Bascule buffers

To protect against human error, a fail-safe mechanism was provided to cut off the supply of high-pressure water to the hydraulic engines just at the end of the operation of

ABOVE The catching arms of a Surrey lower bascule buffer. *(Author)*

smaller as the ram is pushed in, increasing the resistance and decelerating the bascule. Each lower bascule buffer is normally covered with a tarpaulin, to keep water out of the mechanism. When deployed, springs return the buffer to its rest position.

Pawls

Two pawls, situated just below the rear corners of each bascule ballast box, are hinged to brackets bolted to the masonry of each pier. They deploy automatically when the bascule is lowered, preventing traffic on the moving leaf from causing the leaf to move. The two pawls in each pier are withdrawn by the bridge operator before the bascules are raised.

Resting blocks

There are four resting blocks per pier, sitting below the moving girders. Each resting block comprises two inclined blocks, the inclined surfaces in contact with each other and one of the blocks driven by double-acting hydraulic cylinders. In effect, they are wedges which, in conjunction with the pawls and the

raising or lowering the moving leaves. To guard against the failure of this mechanism, hydraulic buffers were installed in each bascule chamber at both high and low level. These were designed to bring the bascule safely to rest, even when the ballast box was travelling at 5ft per second.

The buffers take the form of fixed, oil-filled hydraulic cylinders, fitted with rams and return springs. When struck, oil is forced from the cylinder through openings which become

RIGHT A pawl in the Surrey bascule chamber. *(Author)*

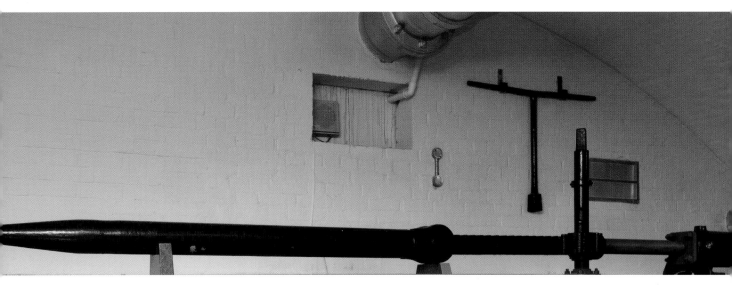

brakes fitted to the hydraulic engines, render the moving leaf 'chocked and locked'.

Nose bolts

At the river end of the Surrey moving leaf, four 5in-diameter steel locking bolts are provided, operated by double-acting hydraulic cylinders positioned directly behind the bolts. The bolts engage in sockets in the Middlesex moving leaf. These bolts act as shear pins,

transferring load between the two moving leaves when the leaves are unequally loaded.

As originally installed, the hydraulic system for the resting blocks and nose bolts was separate from the main high-pressure water supply of the bridge, using a mixture of water and glycerine (anti-freeze) to prevent the fluid from freezing in sub-zero conditions. The hydraulic pumps for this separate pressurised system were driven from the main high-pressure water supply of the bridge. This separation exists to this day.

ABOVE This picture shows one of the original nose bolts with its operating cylinder. Note that the bolts could be operated manually. *(Author)*

LEFT In this photograph the location of the nose bolts can be seen, but there is little to show as bolts and cylinders are hidden from view within the top booms of the main girders of the Surrey moving leaf. The original nose bolts have been replaced, the new bolts being of square section. *(Author)*

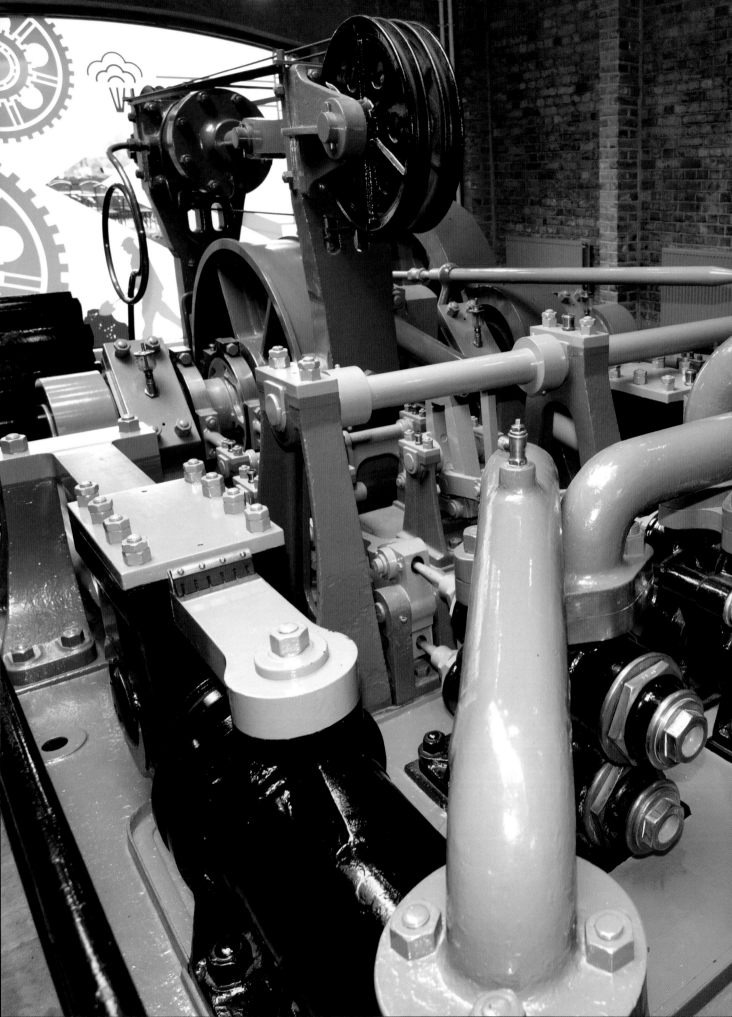

Chapter Six

Bascule bridge operating technology

Operating technology was 1890 state-of-the-art with systems duplicated for high availability. Regulations following the Tay Bridge disaster required the bridge to operate in a wind pressure of 56lb/sq ft. Engines of twice the power needed to operate in this gale were specified, which meant that the bridge machinery had the same tractive effort as 40 Great Western Railway 'King' class steam locomotives.

OPPOSITE The ram end of a large hydraulic engine, which is on display at the Tower Bridge exhibition. Many of the valve gear components can be clearly seen. The weighshaft is the shaft that crosses the engine from side to side, on elevated bearings. This is the only large hydraulic engine remaining at Tower Bridge, the other three having been disposed of. *(Author)*

The overriding principle of the design

Tower Bridge was designed to deliver an extremely high level of availability – defined as the percentage of 24/7 time that the bridge was capable of operating on demand. It may not have had quite the availability of an air traffic control (ATC) system, but it came very close. Everything necessary to operate the bridge was duplicated.

There were two steam-operated hydraulic pump sets and two pairs of boilers, just as an ATC system has two independent electrical supplies and standby diesel generators.

Six high-pressure water accumulators were provided, the equivalent of battery back-up and uninterrupted power supplies in the ATC environment.

There was a duplicated system of high-pressure water distribution, just as an ATC centre has multiple data and voice networks.

There were four engines in each pier, even though one was sufficient to operate a bascule in almost all circumstances, just as an ATC centre has multiple servers, running mirrored

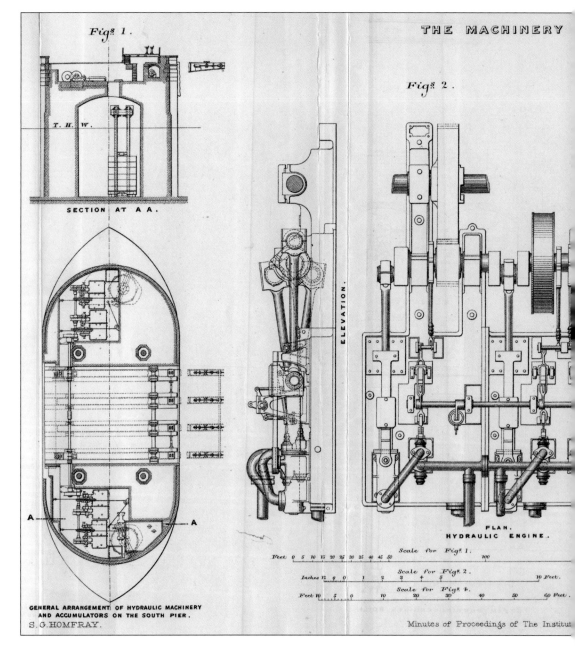

Figs 1.

THE MACHINERY

Figs 2.

T. H. W.

SECTION AT A A.

ELEVATION.

A ——— A

A ——— A

PLAN.
HYDRAULIC ENGINE.

Scale for Figs 1.

Feet 0 5 10 15 20 25 30 35 40 45 50   100

Scale for Figs 2.

Inches 12   9   0   1   2   3   4   5   10 Feet.

Scale for Figs 4.

Feet 10   5   0   10   20   30   40   50   60 Feet.

GENERAL ARRANGEMENT OF HYDRAULIC MACHINERY
AND ACCUMULATORS ON THE SOUTH PIER.
S. G. HOMFRAY.

Minutes of Proceedings of The Institut

databases, ready to take over from each other in an instant.

There were two quadrants on each bascule, two racks on each quadrant, two pinions working on each rack and two pinion shafts in each pier. Even the stairways, lifts and walkways were duplicated. Tower Bridge was designed to be highly available and extremely reliable. In Appendix 2 we will attempt to calculate a percentage availability for the bridge as opened in 1894, well before air traffic control systems had been thought of.

Steam plant – boilers and steam-operated hydraulic pump sets

The working pressure of the Tower Bridge hydraulic machinery was 700 pounds per square inch (psi). This is an enormous pressure, equivalent to the pressure 1,600ft below the surface of the sea and sufficient to crush a Second World War U-boat. Even a modern nuclear attack submarine operating at this depth would be close to its operating limit.

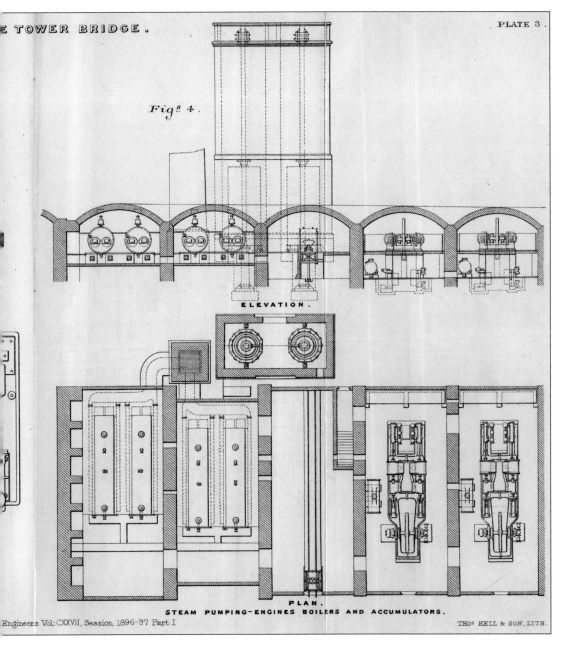

LEFT **Steam pumping engines, boilers, accumulators and hydraulic engines.** *(ICE)*

The layout of the plant rooms under the southern approach is shown on the previous page, taken from the paper presented by Samuel George Homfray to the Institution of Civil Engineers in 1897. Employed by Sir W.G. Armstrong, Mitchell & Co. Ltd, Homfray was responsible for the installation and commissioning of the hydraulic equipment at Tower Bridge.

The hydraulic pressure generation plant occupied five of the arches under the southern approach road, with extensive workshops constructed on the eastern side of the road.

Two of the arches housed the four double-flued, manually stoked Lancashire boilers used for steam generation, two boilers per arch. The boilers are 7ft 6in in diameter and 30ft long, producing saturated steam at a pressure of 85psi. At any time, two boilers were in use, with the other two kept in reserve.

Two arches are occupied by the horizontal tandem steam engines and pumps used to generate the hydraulic pressure for the bridge. Each engine is rated at an indicated horsepower of 360hp and has a high-pressure cylinder 19¾in in diameter and a low-pressure cylinder 37in in diameter on each side of the engine, each low-/high-pressure cylinder pair being coaxial and driven by a single piston rod. Each side of each engine drives a force pump with a diameter of 7¾in and a stroke of 38in. Each tandem engine has sufficient power and pumping capacity to power the hydraulics of the bridge, the other tandem engine and pump set being kept in reserve.

In between the boilers and the engines is an arch occupied by a coal store. The coal was lifted from barges moored against the southern abutment by means of a hydraulic crane and transported to the coal store on small tipping

BELOW The front of one of the Lancashire boilers, together with a pair of steam-operated feed pumps. *(Author)*

LEFT AND BELOW

Here are the flywheel
and pump ends
respectively of one of
the steam-driven pump
sets. The engines
were governed at
15rpm and one engine
was normally ticking
over at 4rpm to top
up the supply of
high-pressure water.
(Author)

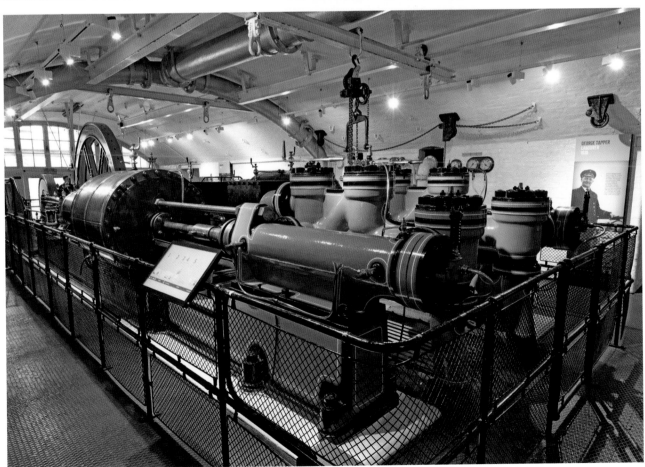

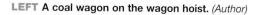

wagons running on rails and pushed by hand. An elevated track running down the centre of the coal store was used to distribute the coal in the store, the wagons being lifted on to the track by a hydraulic hoist.

Hydraulic accumulators

The raised weight hydraulic accumulator is believed to have been invented by Sir William Armstrong; he certainly perfected it. Its purpose was to provide a means of generating and storing high-pressure water for use in hydraulic machinery. Prior to the invention of the accumulator, it was common practice to build very tall towers containing water tanks as a means of generating hydraulic pressure. The tower at the Royal Dock, Grimsby, is a good example. Designed to provide hydraulic power for lock gates and cranes, the water tank was 200ft above ground and contained 30,000 gallons (134 tons) of water. However, the pressure of a 200ft column of water is relatively modest, just 87psi.

A raised weight accumulator does not need to be very tall, yet it can provide pressures of 700–850psi very easily. The principle is very simple. A vertical pipe containing the hydraulic fluid (water in the case of Tower Bridge) is closed at the top by a piston which is weighted. The large volume of pig iron which is normally used to weight an accumulator is held in a large cylinder which is free to move between vertical guide rails.

Water is pumped into the accumulator. When the pressure in the fluid multiplied by the area of the piston equals the weight acting on the piston, the piston begins to rise. Thus, it is simple to design an accumulator to achieve any desired working pressure. When the accumulator is at the top of its stroke, the pumps can be switched off and high-pressure water remains available on demand, the pumps

LEFT **The engine room accumulator tower.** *(Author)*

LEFT A view up into
the engine room
accumulator tower.
(Author)

being used sporadically to top up the volume of high-pressure water available.

At Tower Bridge, there are two accumulators in an accumulator house just to the east of the plant rooms. These have a pipe/piston diameter of 20in and a stroke of 35ft, storing between them up to 153ft^3 (951 gallons) of high-pressure water.

Each pier had two further accumulators, in accumulator chambers at each end of the pier, each with a pipe/piston diameter of 22in and a stroke of 18ft. This enabled each pier to store up to 95ft^3 (592 gallons) of additional high-pressure water.

This begs the question as to whether a bascule could be raised or lowered using purely the stored high-pressure water in the two accumulators in the pier. Given that water is incompressible, it is easy to calculate the maximum volume of water required to lift and lower a bascule using one small hydraulic engine. The circumference of an 82° arc of a circle of radius 42ft is 60.11ft, or 721.3in. The pitch of the rack is 5.9in, so there are 122 teeth on the rack. There are 13 teeth on the pinions which engage with the racks, so 9.4 revolutions of the driving pinions are needed to lift the bascule and another 9.4 revolutions are needed to lower the bascule again, 18.8 revolutions in all. As there is reduction gearing of approximately 6.1:1 between the engine crankshaft and the final pinion driveshaft, the engine crankshaft will need 114.6 rotations to lift and lower the bascule.

Assuming that the engine is operating in full gear with 0% cut-off, the water required will be $3*114.6*(\pi/4)*7.5^2*24in^3$, which equals 364,526in^3, or 211ft^3. So, with the engine in full gear, it would *not* be possible to fully lift and lower the bascule safely using purely the stored high-pressure water in the two accumulators in the pier. However, the drivers could decide how high to lift the bascules, less than a full lift saving high-pressure water. They could also decide how fully to open the valve, which would determine how fast the engine would run. Also, when the bascule was opened or closed, the link blocks in the expansion links were moved automatically towards the centre of the links to cut off the valves before the end of the stroke of the rams, thus economising on water usage, so it is likely that the bascule could have been raised and lowered again using just the water stored in the accumulators in the pier.

The maximum volume of high-pressure water which could be stored in all six accumulators was about 350ft^3. What this tells us is that the bridge could certainly be opened and closed using just the high-pressure water stored in all six accumulators, but one of the steam-driven pumping engines would definitely be operating during a bridge lift to top up the supply of high-pressure water in the accumulators. (Some sources state that the bridge could be opened and closed twice using just the high-pressure water stored in fully charged accumulators.)

One revolution of one pumping engine

pumps 2.075ft^3 of water, so in 1 minute two engines operating at 15rpm could pump 62.25ft^3 of water. This means that it would take only 5 minutes 37 seconds to refill exhausted accumulators.

Pressurised water distribution

The system for distributing high-pressure water to the piers and the engine rooms within them was duplicated, with both manual and hydraulically operated valves installed at many points to allow the system to be configured to operate even when technicians were engaged on the repair or maintenance of part of the system – for example dealing with a leaking flanged joint in a high-pressure pipe, or replacing a seal in an accumulator.

Hydraulic engines

Each pier has two engine rooms. These are positioned just below pavement level at either end of each pier. As originally installed, there were two three-cylinder, single-acting hydraulic engines of unequal power in each engine room. John Wolfe Barry tells us that the design intention was that, in normal

circumstances, one small engine would be capable of lifting a bascule. In a wind pressure of 56 pounds per square foot (unheard of at the location of Tower Bridge), one small and one large engine acting together would be capable of lifting the bascule. This meant that the small and large engine at the other end of each pier would provide a massive power reserve of 100%.

In each engine room, the small and large engines sat side by side, with their crankshafts on the same centreline, the smaller engine being situated towards the end of the pier and the larger engine closer to the centre of the pier. Each engine had a 15-tooth pinion rigidly attached to its crankshaft which engaged with a 41-tooth pinion on an intermediate shaft running behind, and connecting, the two engines. The two 41-tooth pinions were fitted with manually operated dog clutches, enabling one or both engines to be disengaged from the intermediate shaft for maintenance or repair purposes. A 13-tooth pinion on the end of the intermediate shaft closest to the bascule engaged with a 29-tooth pinion on the final pinion driveshaft, effecting an overall step-down gearing of approximately 6.1:1 between the engine crankshafts and the pinion driveshaft. The small engines have a stroke of 24in and cylinders 7½in in diameter. The large engines have a stroke of 27in and rams 8½in in diameter. Separate piston valves are provided to admit and exhaust high-pressure water to and from each ram. These are similar to the valves in a cornet. They are operated by 'pokers' and closed by return springs. The valve and reversing gear is a simplified form of Stephenson valve gear. Stephenson gear normally features a forward eccentric and a backing eccentric, each driving, by means of an eccentric rod, one end of an expansion link which can be lowered and raised to change the position of the link block working within the link. In the Tower Bridge hydraulic engines, each expansion link is pivoted near its centre and driven by a single eccentric rod connected to the lower end of the link. The blocks, rather than the links, are raised or lowered by the lifting links attached to the weighshaft lifting arms.

Rods transfer the motion of the blocks to rocking levers which actuate the 'pokers'. The position of the blocks in the expansion links determines the direction of the engine and the stroke given to the valves. The starting position (full forward or full reverse) was determined by the initial movement of the main starting valve (or regulating valve) in the control cabin, the reversing gear being operated by a hydraulic cylinder. A screw and nut arrangement on the end of the pinion driveshaft moved the weighshaft automatically to cut the supply of high-pressure water to the engines as the bascule approached the end of its intended movement.

Tuit tells us that the total force which must be applied to the quadrant racks in a wind

BELOW This is the crankshaft end of the same large hydraulic engine shown on page 115. A section of the intermediate shaft and one of the dog clutches is clearly visible. The braking system is a prominent feature. A large brake wheel is fitted to the crankshaft against which two brake blocks work. When the engine is required to operate, the brake blocks are forced away from the brake wheel by a hydraulic cylinder. When the brake needs to be applied, or if hydraulic pressure is lost, the brake blocks are forced against the brake wheel by means of a heavy weight attached to the end of a steel wire wound around the two high-level grooved wheels on the brake gear. *(Author)*

exerting a pressure of 56 pounds per square foot on a moving leaf is about 190 tons. Let's see if he is right.

The Tractive Effort (TE) of a steam locomotive with two double-acting cylinders is usually calculated as:

TE = $0.85*d^2*p*s/w$ pounds, where 0.85 is a coefficient which reflects the fact that boiler pressure in practice might be 85% of rated pressure; d is the piston diameter in inches; p is the boiler pressure in pounds per square inch; s is the stroke in inches; and w is the diameter of the driving wheel in inches. To be accurate, there should also be a multiplier of $\pi/4$, but this is normally overlooked, and we will do the same.

The formula shows that the smaller the driving wheels of a locomotive, the larger the tractive effort. Tractive effort was a very 'big deal' for the Great Western Railway (GWR), London, Midland & Scottish Railway (LMS), London & North Eastern Railway (LNER) and the Southern Railway (SR) in the heyday of the steam locomotive. Every railway wanted to hold the blue riband for the most powerful locomotive. When Chief Mechanical Engineer Charles Benjamin Collett designed the 'King' class locomotive for the GWR in 1926, he went to the substantial expense of specifying 6ft 6in driving wheels, rather than the customary 6ft 8½in wheels used on all GWR express locomotives, purely to wrest the blue riband back to the GWR from the SR's 'Lord Nelson' class.

For hydraulic engines, the pressure is fixed, so we can remove the coefficient. As the Tower Bridge hydraulic engines were single-acting, we must divide by two, but as the engines had three cylinders we must multiply by 1.5. Therefore, the formula for the tractive effort of the Tower Bridge engines is:

TE = $0.75*d^2*p*s/w$ lb

w is the pitch circle diameter of the final 13-tooth pinions driving the racks, which is 24.5in. The reduction gearing between the engine crankshafts and the final pinion drive shaft is 6.097, so:

TE = $6.097*0.75*d^2*700*s/24.5 = 130.65*d^2s$ lb

So, the tractive effort of a small hydraulic engine is 176,377.5lb (78.74 tons), and the tractive effort of a large hydraulic engine is 254,865.5lb (113.78 tons), making the total tractive effort of a small and large engine working together equal to 431,243lb (192.5 tons) – almost exactly the Tuit figure. Incidentally, this is more than the tractive effort of ten of Mr Collett's 'King' class locomotives. And there was the power of another ten 'King' class locomotives on call at the other end of the pier. The Tower Bridge hydraulic engines were not short of power!

Bridge control

Bridge control and interlocking equipment was provided by Messrs Saxby & Farmer, well-established signalling contractors who supplied signalling equipment and signal boxes to many British railway companies. Two lever frames exactly like those in railway signal boxes were provided to control the hydraulic machinery, together with mechanical interlocking frames below the floor of each control cabin, also not dissimilar to those in a railway signal box. The interlocking frames were designed to prevent, for example, any hydraulic engine from being operable until the pawls and resting blocks had been withdrawn. The nose (or locking) bolts were controlled from the south control cabin only.

A control, or driver's, cabin was built on the east side of each pier and a watchman's cabin constructed on the west side of each pier. Stairs in each cabin led down to the engine room below. The cabins were constructed of wood, which would have been perfectly normal for Saxby & Farmer, but seems a little bizarre today, given the majestic construction of the bridge itself. They were rebuilt more sympathetically, in stone, after the Second World War.

Semaphore signals, indicating to shipping during daytime whether the bridge was open or shut, were mounted on top of the cabins, operated manually by levers in the cabins. These were, perhaps, a little superfluous, as the bascules themselves provided the largest semaphore signal in the country. Four red and four green lights in both upstream and downstream directions performed a similar function at night.

It is not clear whether corresponding signals for road traffic were initially provided. Tuit, writing before the bridge had been opened to traffic, described the proposed procedure, which involved policemen and chains across the road within the archways of the main towers. The chain would be fixed to the wall at one side of the archway. A policeman would stop the traffic and the free end of the chain would be placed over a hook on the other side of the archway. The person undertaking this task would then turn a valve and a small hydraulic cylinder would draw the chain tight and also connect with the locking frame under the control cabin, such that the chain had to be in place, and tight, before the moving leaf could be moved.

Small additional Saxby & Farmer interlocking lever frames were provided some years after the bridge was commissioned to control road and river signalling in a more convincing manner, fully interlocked with the bridge controls. This is unlikely to have been as late as the early 1950s, following the infamous incident on 30 December 1952, in which a number 78 double-decker bus was crossing the bridge when a bascule started

to move. So this incident really should have been prevented!

Hydraulic lifts

The lifts in the main towers were each operated by duplicated hydraulic cylinders. Lifting ropes and counterweight ropes were duplicated also, and interlocking gear prevented movement of a lift until both inner and outer doors were closed.

Fire, domestic water and bilge hydraulic pumps

Hydraulic pumps in the engine rooms on the piers were provided to deliver water to the tops of the main towers in the event of a fire. Other hydraulic pumps also provided water to the accommodation in each abutment tower.

Hydraulic bilge pumps were installed in the accumulator chambers on each pier to deal with any seepage from the accumulators.

RIGHT This is the underside of the Saxby & Farmer frame in the south control cabin. The counterweights relieved the driver of some of the weight of the various vertical control rods linking with the hydraulic engines. *(Author)*

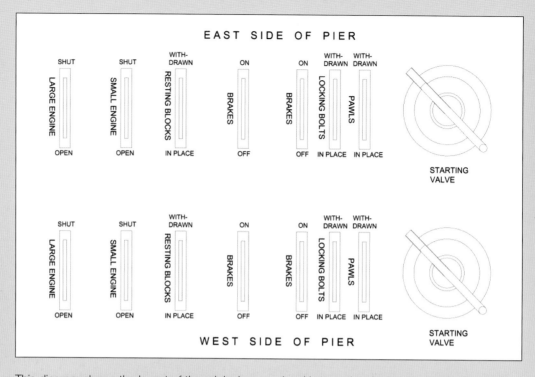

This diagram shows the layout of the original controls in the south control cabin. Two rows of levers were provided, one for use when the machinery on the east side of the pier was to be deployed and one for operating the engines on the west side of the pier. The levers were operated like point levers in railway signal boxes; they had to be in one extreme position or the other, not at some point in between. At the end of each row of levers was a large spindle valve, variously known as the starting valve or the regulating valve. Homfray describes, rather arcanely, that the first movement of the starting valve handle 'against the sun' would cause the hydraulic cylinder controlling the engine weighshafts to set the valve gear to lift the bascule. Further movement opened the regulating valve, the extent to which the valve was opened determining the speed of the selected engine. Movement 'with the sun' back to the original position would cut off the supply of high-pressure water to the engine, with further movement first fully reversing the engine and then supplying it with high-pressure water. Either or both hydraulic engines in the chosen engine room could be selected, and brake levers were provided for both engines.

The picture opposite (top) shows the inside of this cabin today, still containing its original equipment. The bank of gauges on the wall (opposite) showed the driver the pressure available to operate the machinery from the two separate high-pressure water supplies (A and B), plus the pressure in the return/exhaust pipe. They give tangible evidence that the nose bolts are either locked or unlocked, and show the

BELOW The tell-tale board in the south control cabin.

pressure admitted to the chosen engine(s) by means of the regulator valve.

Above the water pressure gauges is a tell-tale board showing the position of the Surrey resting blocks and pawls, the nose bolts and the road signals at both ends of the bridge.

A telephone system for communicating between control and watch cabins, and between control cabins and the engine room under the southern approach was installed by Messrs Spagnoletti and Crookes. There was also a system of signalling bells and a set of bell codes, again inspired by railway signalling practice.

ABOVE Original bridge controls in the south control cabin.

LEFT Water pressure gauges in the south control cabin.

BRIDGE OPERATING PROCEDURE

The procedure to operate the bridge was as follows. The Bridge Master would have decided which engine room on each pier was to be used that day and the engine drivers would have selected either the large engine or the small engine in the chosen engine room by means of the levers. Clutches would not normally be operated on a daily basis, the three unused engines being kept in gear and allowed to run idle. As the bridge was required by Act of Parliament to lift on demand, a watchman would be on lookout at all times in all four cabins and a driver would be available in each control cabin. The approach of a vessel would be signalled by a bell code. The Head Watchman would decide when to instruct the clearance of the bridge. On his order, the City of London policeman on duty on each pier would operate the traffic signals and clear traffic from the moving leaves. Watchmen would clear foot passengers from the moving leaves and close barriers across the roadway and the pavements.

1 On the Head Watchman's signal, the drivers would retract the pawls and resting blocks and the driver on the Surrey pier would withdraw the nose bolts.

2 3 The driver on the Middlesex pier had no way of knowing when the nose bolts had been withdrawn, so the Surrey bascule always lifted first, the driver in the Middlesex control cabin operating his starting valve when he saw the first movement of the Surrey bascule.

4 5 Once the bridge was as open as necessary, the semaphore signals or lights for shipping were changed to advise the master of the approaching vessel that he could proceed. The drivers would apply the brakes on the engines being used. Normally, the moving leaves were not opened fully, this being reserved for 'a salute of honour'.

6 The procedure was essentially reversed for lowering the bascules.

Tower bridge.

*(Ian Moores –
ianmooresgraphics.com)*

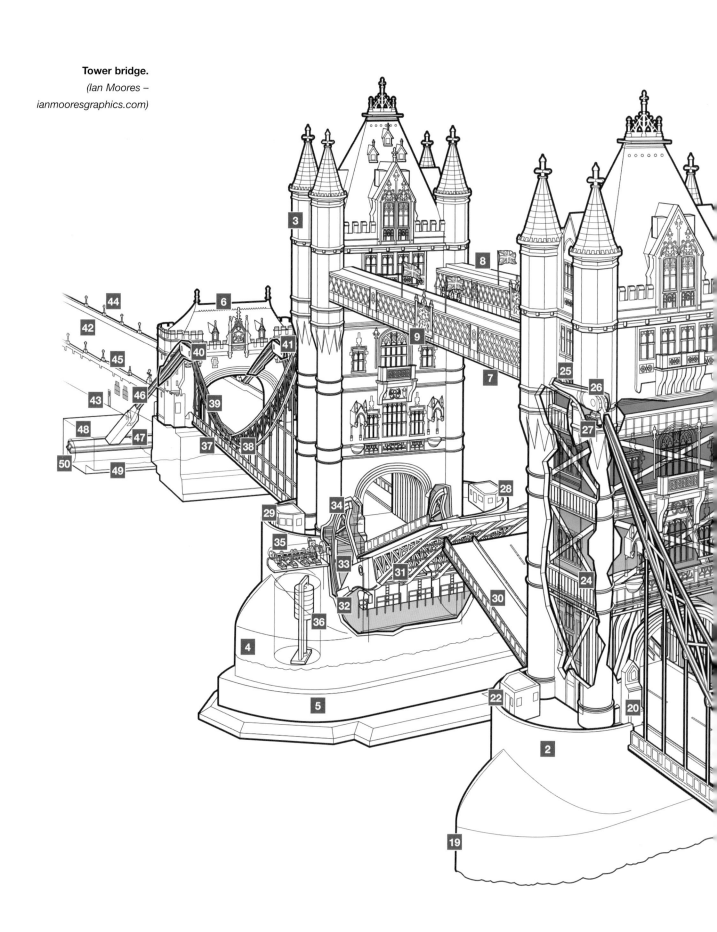

1 Surrey (or south) main tower
2 Surrey pier
3 Middlesex (or north) main tower
4 Middlesex pier
5 Concrete foundation of Middlesex pier
6 Middlesex abutment tower
7 West high-level walkway
8 East high-level walkway
9 Arms of the City of London
10 East Surrey long chain
11 Top boom of east Surrey long chain
12 Bottom boom of east Surrey long chain
13 West Surrey long chain
14 Surrey shore span
15 Cast iron bridge parapet
16 Pinned joint connecting suspender rod to long chain
17 Suspender rods
18 Pinned joint connecting west Surrey long and short chains and the west Surrey stiffening girder

19 Cutwater
20 Masonry projection to accommodate the west Surrey bascule quadrant and racks
21 Surrey pier control or driver's cabin
22 Surrey pier watchman's cabin
23 One of the two A-frames supporting the steel arch below the Surrey main tower
24 A wind brace
25 The west high-level tie
26 The pinned joint between the west Surrey long chain and the west high-level tie
27 The rockers transferring suspension bridge load to the west landward steel column of the Surrey main tower
28 Middlesex pier control or driver's cabin
29 Middlesex pier watchman's cabin
30 Surrey moving leaf
31 One of the four main girders of the Middlesex moving leaf
32 The ballast box of the Middlesex bascule
33 The position of the main pivot shaft
34 The west quadrant of the Middlesex bascule, to which two separate racks are bolted, driven by four pinions, two on each of the two pinion shafts
35 The west Middlesex engine room
36 The west Middlesex hydraulic accumulator
37 The Middlesex shore span
38 The west Middlesex long chain
39 The west Middlesex short chain
40 The west Middlesex horizontal link
41 The east Middlesex horizontal link
42 The northern approach viaduct
43 Accommodation beneath the arches of the northern approach viaduct for the Tower of London
44 Cast iron lamp standard
45 The inconspicuous chimney of the Guard Room
46 The west Middlesex land tie
47 The west Middlesex anchor tie
48 The west Middlesex anchor girder
49 The Middlesex concrete anchor block
50 The inspection subway running through the Middlesex anchor block

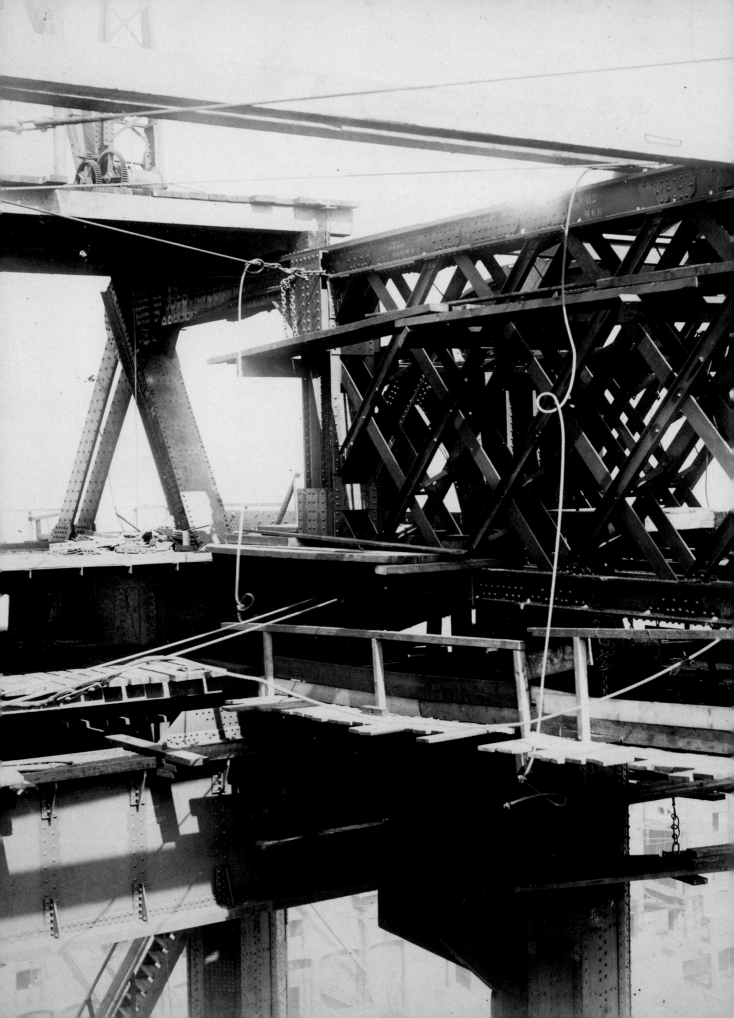

Chapter Seven

The high-level bridges

Few visitors to the bridge will notice that each footbridge comprises a central section suspended by forged links from cantilevered end sections. The Achilles heel of Tower Bridge is that 156 tons of high-level tie weight must be supported – the weight of 2,500 pedestrians. Originally, this weight was carried by the outer footway girders. Since 1960, the weight has been carried by catenary wire suspension bridges.

OPPOSITE Work in progress. The wooden platforms, slung by ropes from the girders, prevented rivets, bolts, small tools and larger items from falling on to vessels passing below. They also provided the means for the workmen to undertake the riveting of booms, ties and struts. Today, the flimsy and open wooden handrails would be seen as a health and safety disaster waiting to happen. However, Cruttwell writes, with obvious satisfaction: 'It is gratifying to note that during the eight years occupied in building the bridge, out of an average number of 432 men employed during the whole of that period, there were only ten fatal accidents.' That's nearly 2.5% of the workforce dead, George! *(LMA/Collage Image 323337)*

The high-level footbridges

Overview

There are actually four high-level bridges. These comprise the two original high-level walkways and the two catenary wire suspension bridges which now carry the dead weight of the high-level ties of the suspension bridge, relieving the walkways of this burden. The purpose of the walkways was twofold.

Firstly, they were designed to provide a means for foot passengers to cross the bridge when the bascules were raised. When first completed in 1894, the bascules were raised about 17 times per day. The following year there was a small increase, the bascules being raised about 20 times per day. Therefore, interruption of vehicular and foot passenger traffic was frequent. It could also be prolonged, if several vessels followed each other into, or out of, the Inner Pool. The average interruption to traffic was 6 minutes and the longest interruption to traffic during the first two years of operation

LEFT One of the four circular staircases between roadway level and the first landing. *(Author)*

BELOW LEFT A stairway further up a tower where, free from the constraints imposed by the bascule quadrants and the arch, the stairs were implemented as straight flights. *(Author)*

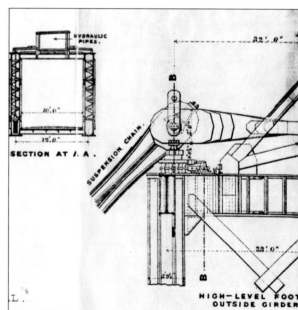

was 36 minutes. Two sets of stairs (one up, one down) in both main towers were accessible to the public between sunrise and sunset for passengers in a hurry, but there were 206 steps up to the walkways and the same number down on the other side. These staircases have changed very little.

Two hydraulically operated 'hoists' (lifts/elevators in modern parlance) were also installed in each tower, capable of taking 25 passengers to and from the footway level with a maximum waiting time of 2 minutes 30 seconds. However, the Bridge Master decided not to open these to the public, as the average interruption to road and foot passengers was less than the time taken to wait for and take the lift, cross the footbridge and wait for and take the lift down in the other tower. This reduced substantially the quantity of high-pressure water and coal needed to operate the bridge.

The secondary role of the footbridges was to support, and provide protection for, the two high-level ties of the suspension bridges, which are positioned at low level inside the outer girder of each footway.

Each footbridge, footway or walkway consists of a simply supported central section slung between two cantilevered end sections, one attached to the Middlesex superstructure and the other to the steelwork of the Surrey main tower.

Each cantilevered section comprises an outer cantilever, riveted to a section of outer footway girder, and an inner cantilever, riveted to a section of inner footway girder. The top booms of the outer and inner footway girders are connected by fabricated I-beams, spaced at intervals of approximately 6ft, and by wind bracing, as are the bottom booms.

Near the ends of the top booms of each cantilevered section of footway, 5in-diameter pins carry forged steel links which connect to similar 5in diameter pins in the vertical ends of the girders of the central section. Inner and outer girders are 12ft 10in high and feature a double lattice of flat bars for the ties and channel bars for the struts.

The centres of the inner and outer girders are 12ft apart, giving a clear width of 10ft for foot passengers.

The cantilevered footway sections

John Wolfe Barry tells us in his lecture that the cantilevered footway sections extend 55ft from each tower. Cruttwell tells us, more precisely, that the distance from the centres of the river-facing main tower columns to the vertical ends of the cantilevered footway sections is 58ft 6in, leaving a space of 120ft for the central section of footway. However, the drawing attached to his paper shows the distance from the centres of the river-facing main tower columns to the vertical ends of the cantilevered footway

BELOW This general arrangement is taken from Cruttwell's paper on the superstructure of the bridge. An outer cantilever is shown on the left of the diagram and an inner cantilever on the right. *(ICE)*

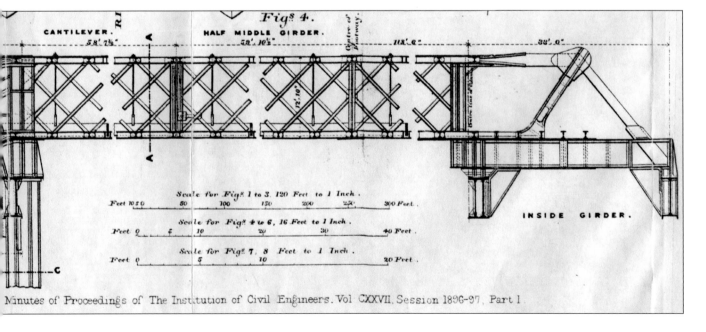

Minutes of Proceedings of The Institution of Civil Engineers. Vol CXXVII, Session 1896-97, Part 1

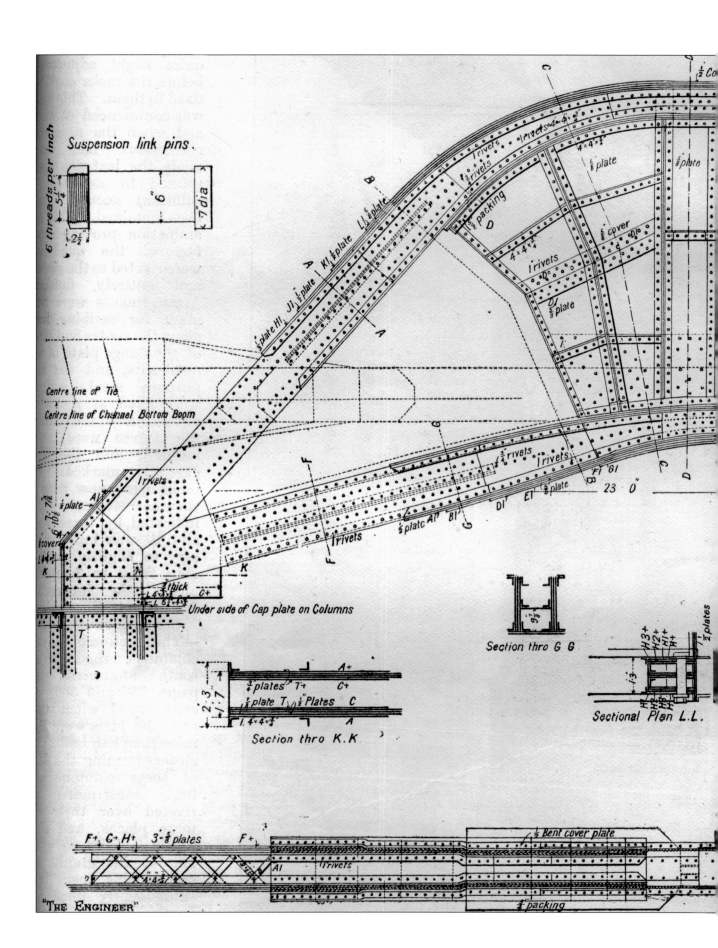

Suspension link pins.

6 threads per inch

7" dia

6"

5 1/4"

2 1/2"

Centre line of Tie

Centre line of Channel Bottom Boom

Under side of Cap plate on Columns

Section thro K.K.

Section thro G G

Sectional Plan L.L.

"THE ENGINEER"

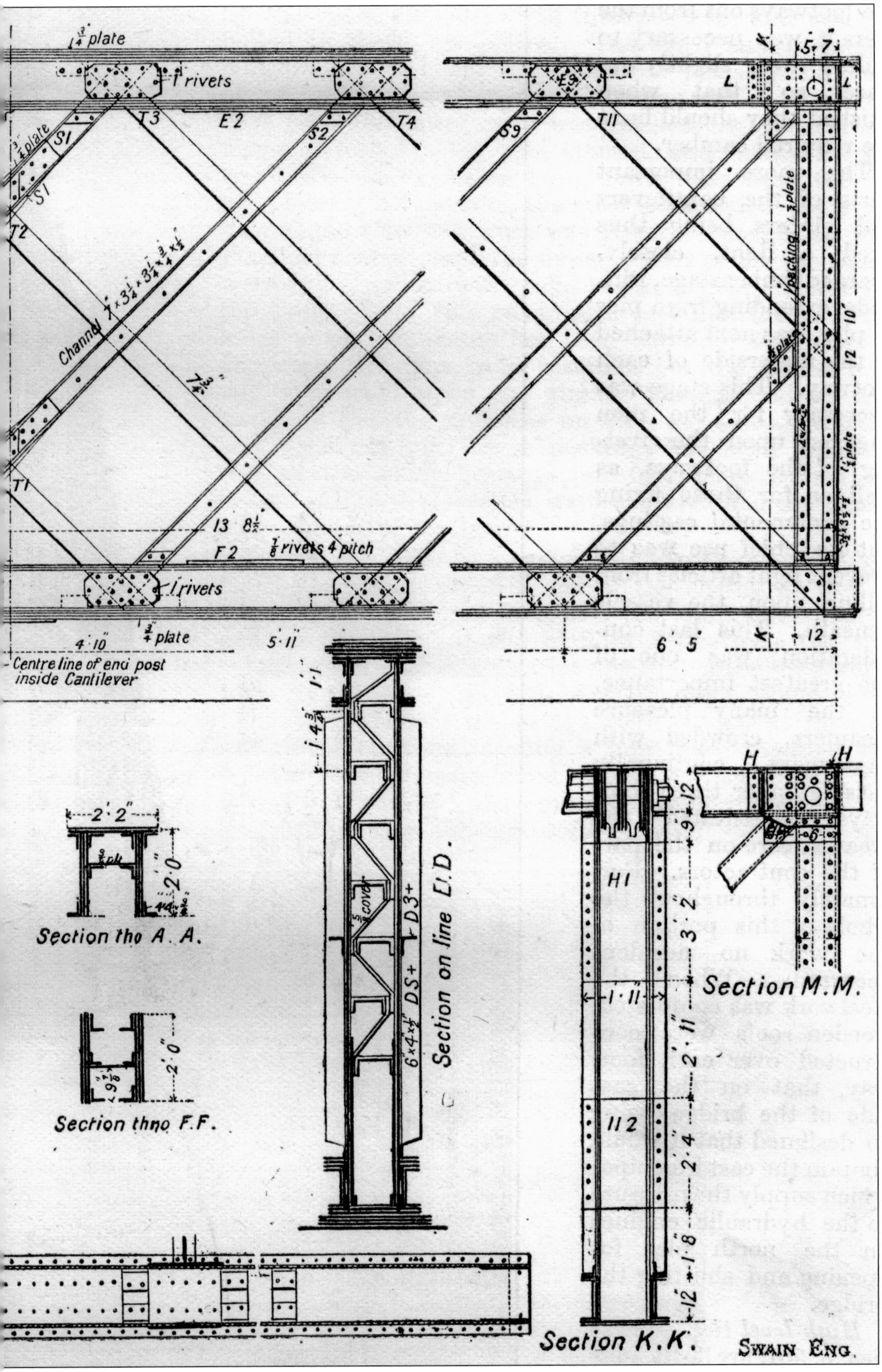
plate
rivets
T3
E2
S2
T4
E9
T11
S9
S1
T2
Channel
T1
F2
rivets 4 pitch
plate
4·10"
5·11"
6·5
Centre line of end post inside Cantilever
packing
plate
H1
H2
Section thro A. A.
Section thro F.F.
Section on line L.D.
Section M.M.
Section K.K.
Swain Eng.

LEFT An outer cantilever with its integral section of outer footway girder, taken from Tuit's paper. The quality of reproduction is not high, but the drawing shows the key features. We see that the high-level tie sits within the lower boom of the outer lattice girder and that the upper arm of the cantilever (the member which is in tension) is divided into two to allow the tie to pass through it. *(ICE)*

RIGHT The east inner cantilever of the Middlesex main tower. This is one of the author's favourite parts of the bridge and is a wonderful example of 'form ever following function'. It would make a formidable opponent for a tug-of-war team. *(Author)*

sections as 58ft 7½in. Tuit has yet another take, advising that the cantilevered sections extend 59ft beyond the centres of the columns, and that the central footway sections are 118ft 9in long, which leaves 1½in of space at each end of the central sections. There is clearly some confusion here!

The lower member of the cantilever (in compression) converges with the upper member and the two are riveted to each other and to a main third landing girder of the tower, a little way from the rockers under the end of the high-level tie. The outer girder passes through the stanchion atop the river-facing column and the weight of the outer cantilever is borne by this column. We also see the hole for the suspension pin in the top boom of the outer cantilevered girder.

The inner cantilever is securely attached to one of the main cross girders, which sit on top

LEFT This view is taken from the east Surrey stairway just below the third landing. At this point the stairway passes directly below the cross girder carrying the east inner cantilever. The two gussets connecting the cross girder to the south third landing girder and helping to carry the tensile load of the cantilever can be seen, their design complicated by the through passage of a wind tie.

The upper and lower flanges of the south third landing girder can also be seen, the space between them filled with brickwork. We also see the first layer of stonework on top of the third landing girder. The brickwork below the third landing girder finishes just below the bottom flange of the girder. This tells us that no masonry loads are transmitted between landings. *(Author)*

of the north and south third landing girders in each tower. This particular cross girder extends a little beyond the main landing girder to provide direct support for the end of the inner cantilevered footway girder.

The simply supported central footway sections

Each of the two central footway sections comprises an inner girder and an outer girder. Roof and floor I-beams, wind bracing, floor plating and roofing is identical to that on the cantilevered sections of footway.

Roofing and ornamentation

The roofs of the footways are very slightly pitched, so that they shed water. As originally built, the roofs were of wooden construction with a sheet zinc covering. The roof of the east footway featured a central raised section protecting two 6in pipes supplying the Middlesex tower with high-pressure water, one 7in pipe carrying water exhausted from the Middlesex tower hydraulic engines back to the reservoir under the southern approach, hot water pipes to prevent water in these pipes from freezing, plus other water, electricity, gas and telephone services. These services were run up the centre of the eastern river-facing column of the Surrey tower and down the centre of the eastern river-facing column of the Middlesex tower.

Cast-iron ornamental panelling was applied to outer and inner footway girders, occupying approximately one-third of the height of the girders. Large ornamental cast-iron castings showing the coat of arms of the City were applied to the centres of both sides of the two suspended central footway sections. Decorative iron grilles were applied to the square spaces between the ties and struts of the footway girders. These were unglazed, making the walkways draughty and cold in winter. The decorative tracery did prevent people from climbing out of the walkways.

This ornamentation is visible in the photograph on page 77, but much of it had been removed by the mid-1940s, as the cast-iron panels prevented effective maintenance of the lower portions of the footway girders. In the late 1970s, steel mesh maintenance walkways (supported by short sections of rolled steel joist welded to the lower booms of the outer girders)

RIGHT The junction of a cantilevered footway section with its central section. The hex nuts, 4½in across flats, which hold the suspension pins in place are clearly visible. *(Author)*

ABOVE **Here in 1894 is the inside of a footway as built, possibly prior to the installation of the floor covering. The slanting wooden dado panelling and the gas lamps give a pleasant, if rather austere, ambience.** *(ICE)*

were installed on both sides of each footway. Glass reinforced plastic (GRP) panelling was added to reproduce the look and feel of the original cast-iron panels. New aluminium double glazing was installed between footway girder ties and struts, and new aluminium roofing installed, the large services duct on the top of the roof of the eastern footway no longer being required.

The footways have been subject to substantial change, including work concentrating on the removal of excess weight, weather-proofing, the provision of improved facilities for maintenance and the creation of entertainment spaces.

Method of construction

Today we build aircraft carriers in sections – complete with services, machinery and accommodation – and weld them together. If we were building the Tower Bridge footways today, we would definitely build two complete sections of central footway, complete with roofing and flooring, and crane them up to be suspended from the completed cantilevers. This was not seen as an option in the 1890s. Cranes capable of performing this 200-ton lift of 140ft were not available and the Tower Bridge Act prohibited the contractor from preventing the passage of shipping.

The only alternative was to build the footways out from each tower as cantilevered sections, including the inner and outer girders of the central, simply supported, sections of footway, until they met in the centre of the river, some 115ft from the face of the main towers. Work progressed on all four cantilevered footway sections simultaneously. This provided fascinating viewing for Londoners and rather more adrenaline than was good for them for Sir William Arrol's workmen.

When the cantilevered sections of footway had been completed, the vertical ends of the central footway section girders were attached, slung by the pins and links. The first sections of central section bottom booms were then added, followed by the first few diagonal struts and ties and the first sections of top boom. Supporting these first sections of central footway girders by cranes, iron blocks were placed between the vertical ends of the bottom booms of the cantilevered footway sections and those of the central girders. Temporary ¾in-thick steel joint plates, 30in tall and 12ft long, were then attached to the inner and outer girders, straddling the joint between the cantilevered and central sections of footway, this providing sufficient support for the construction of the central footway girders to proceed out across the river.

As a complete set of parts for the four central section girders had been fabricated and trial-assembled at the Dalmarnock Iron Works, it was necessary to ensure that the correct space was provided between the verticals of the cantilevered and central girders such that the last pieces of central section booms, ties and struts would fit perfectly. This necessitated the precise measurement of the distance between the vertical ends of the cantilevers using a steel wire stretched across the river.

One wonders whether it might not have been simpler and less risky to have made the last sections of the central footway girders in a bespoke manner 'to fit the job'. In the event, Arrol's strategy worked, it only being necessary to wait until the steel of the girders was at the right temperature to allow the last few 'pieces of Meccano' to be slotted in and riveted. The cast-iron blocks were then knocked out and the temporary steel joint plates removed, leaving the central sections of footway simply supported at their ends – as they remain, 125 years later.

High-level tie support

Each high-level tie is 301ft long between the centres of the holes in the eyes, and each tie weighs 120 tons. I was keen to know whether the ties, in the absence of support along their length, would be self-supporting and what the deflection at the centre would be. This is easy to establish. The calculation shows that, if the tie was sitting on supports under the holes of the eyes and not subjected to any tension, the self-weight of the tie would cause it to sag by 51ft in the centre, and the ends would be at an angle of 31° to the horizontal. The bending stress in the centre of the tie would be 70.5 tons per square inch. It could not survive that level of stress. If the tie were under its normal tension of around 900 tons, it would probably be capable of supporting itself, but I would hate to see this put to the test. Clearly, supporting the dead weight of the ties is vital.

Each tie was supported by 1½in-diameter rods, suspended from the top boom of the outer footway girder or, in the case of the suspender rod nearest to the end of each tie, from the upper boom of the outer cantilever. At their lower ends, the suspender rods attached to a cradle bolted to the top of the tie by means of one of the 1in fitted bolts in the top row of bolts connecting the two halves of the tie together. The bottom edge of the high-level tie was about 6in from the 'floor' of the bottom boom. The suspender rods were spaced at intervals of about 12ft. Each cantilevered section of footway had five suspender rods,

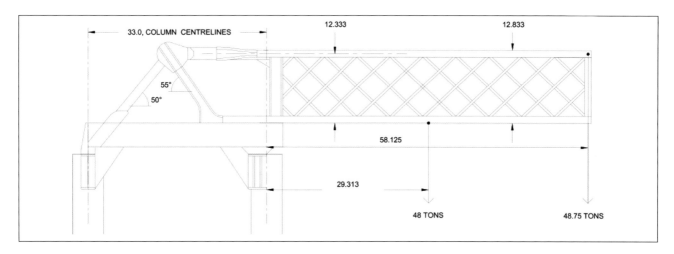

ABOVE **Force diagram for an inner fixed footway girder, as originally constructed in 1894.** *(Author)*

supporting between them about 19.5 tons of the weight of the high-level tie. Each middle section of footway had ten suspender rods, supporting between them about 39 tons of the weight of the high-level tie.

Forces in the structure

The Schedule of Quantities in Contract No. 6 (superstructure) gives the weight of steel in the cantilevers and the inner and outer footway girders as 457 tons. Estimating the weight of each outer cantilever as 15 tons and that of each inner cantilever as 10 tons, this leaves 357 tons for the steel in the fixed and central footway girders. The Schedule of Quantities also quotes the weight of steel in wind-bracing, floor plates and so on as 130 tons, giving a total weight of steel in the footways of 487 tons. Knowing the length of the fixed and central sections of footway, the weight of a complete, simply supported

middle section of footway can be estimated. The east footway was the heavier of the two, due to the services duct on its roof. The table below shows the estimated weight of the east central section of footway as approximately 195 tons (excluding the weight of the high-level tie). From this figure, the estimated weight of each east cantilevered section of footway (excluding the weight of the cantilevers themselves and the weight of high-level tie supported) is 96 tons.

From these figures, we can draw a simple force diagram for an inner fixed footway girder, as originally constructed in 1894 (above). Dimensions are in feet and weights are in tons.

Half the weight of the cantilevered section of footway acts halfway along the inner girder, and quarter of the weight of the middle section of footway acts at the end of the inner girder. None of the weight of the high-level tie is carried by the inner girder.

Weight of the east walkway middle section					
					Weight
Steel					123.03
Cast-iron panels (estimated at 157 tons in total)					39.66
Tracery (total weight is 56 tons)					14.15
Wood in roof = 14ft wide × 119.75ft long × 2in thick					4.37
Wood in floor and internal dado panels					2.76
Wood in services duct 10ft × 119.75ft × 2in					3.12
	Area in^2	**Length** in	**Volume** in^3	**Density** lb/ft^3	
6in pipe	10.21	2,874	29,344.0	485	3.68
7in pipe	11.78	1,437	16,929.3	485	2.12
3in pipe	5.50	1,437	7,900.3	485	0.99
Other services (estimate)					1.00
				Total	**194.89**

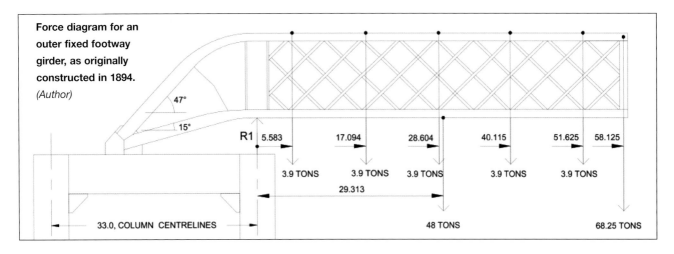

Force diagram for an outer fixed footway girder, as originally constructed in 1894. *(Author)*

47°
15°

R1 | 5.583 | 17.094 | 28.604 | 40.115 | 51.625 | 58.125

3.9 TONS | 3.9 TONS | 3.9 TONS | 3.9 TONS | 3.9 TONS

29.313

33.0, COLUMN CENTRELINES | 48 TONS | 68.25 TONS

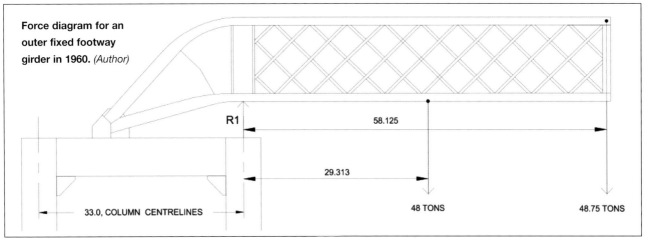

Force diagram for an outer fixed footway girder in 1960. *(Author)*

R1 | 58.125

29.313

33.0, COLUMN CENTRELINES | 48 TONS | 48.75 TONS

An analysis of the forces in the structure shows that:

■ The horizontal pull exerted by the inner cantilever is 287.8 tons.
■ The tension in the shore-side angled member is 167.8 tons.
■ The compressive force in the river-side angled member is 156.9 tons.
■ The vertical weight acting on the river-side column is 225.3 tons.
■ The shore-side column experiences a vertical pull of 128.5 tons.
■ The inner fixed footway structure imposes a clockwise moment of 5,837.0 tons-feet about a point at the bottom of the foundation of the pier and on the pier centreline.

The outer fixed footway structure imposes a clockwise moment of 8,171.7 tons-feet about a point at the bottom of the pier foundation and on the pier centreline.

So, the net clockwise moment in 1894 about a point at the bottom of the foundation of the pier and on the pier centreline due to both footways is 2*(5837.0 + 8171.7) tons-feet = 28,017.4 tons-feet.

In about 1960, high-level tie suspension bridges were installed, relieving the outer footway girders of the weight of the ties. The lower of the two images above is a force diagram of an outer fixed footway girder in 1960. The structure now imposes a clockwise moment of 5,837.0 tons-feet about the centre of the pier foundation (*ie* the same as an inner footway girder). So, the net clockwise moment in 1960 about a point at the bottom of the foundation and on the pier centreline due to both footways is 2*(5837.0 + 5837.0) tons-feet = 23,348.0 tons-feet. This is probably overstated by some 15%, as over 200 tons of cast iron had been removed from the footways by the mid-1940s. However, the work undertaken to prepare the footways for public access in the early 1980s, which included glazing, will have added back some of this weight.

RIGHT A footway in
1946. The steelwork
is totally open to
the elements. This
is definitely the low
point in the life of
the footways! (Getty
Images 3360497)

BELOW The
connection of a pair of
suspender rods with
their carrier. Adjusters
provide the means
to ensure that the
suspended high-level
tie is straight and level.
(Author)

The high-level tie suspension bridges

By the end of the Second World War, all the cast-iron panels decorating the high-level walkways had been removed in order to reduce the weight carried by the footway girders and to provide access to the lower parts of the girders for inspection and

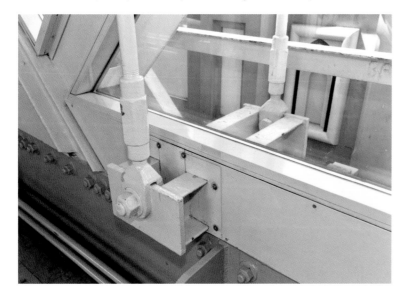

maintenance purposes. However, corrosion of the high-level walkways continued to cause concern. In the mid-1940s, lattice cross girders were installed in an attempt to make the inner footway girders share some of the load of the high-level ties. This was not a very satisfactory solution as the weight of the high-level ties was still being carried by the footways and the steel cross girders added substantial *extra* weight. More importantly, the footways had been rendered incapable of fulfilling their original purpose.

Finally, in 1960, a permanent and effective means of relieving the outer footway girders of the weight of the high-level ties was installed in the form of two high-level tie 'suspension bridges'. These transfer the dead weight of the high-level ties to the superstructure of the main towers. It was a neat and effective solution and one which is probably unnoticed by most tourists visiting the bridge, the internal catenary wire looking very much like a handrail.

Design
Occasional cross bracing and the links suspending the middle footway sections from

the cantilevered footway sections prevent a single catenary cable solution, so two cables were used to support each high-level tie, one inside the footway and one outside. The load of the tie is carried by 12 pairs of suspender rods. These are spaced approximately every 12ft along the footway. Each suspender rod terminates in a cable clamp at its upper end and holds one end of a carrier at its lower end.

The use of two cables fits neatly with the high-availability design ethos of the bridge and, of course, if either cable were to fail, the tie would simply sit down on the floor of the bottom boom of the simply supported middle section of footway. This would not increase the tension in the outer cantilevers beyond their original design level, but it would severely stress the two suspenders attached to the upper booms of the outer cantilevers, if these suspenders are still in place.

Cable clamps, suspenders and carriers

The carrier is fabricated from two lengths of steel channel welded to end plates to which eyes on the suspender rods are bolted. In between the lengths of channel is the boss of a short suspender rod carrying the weight of the high-level tie. It is unclear whether this rod is bolted, screwed or welded to the high-level tie.

BELOW A simple force diagram for a single high-level tie suspension bridge, comprising two catenary cables. The angle at which the cables meet the superstructure of the towers is estimated at 12°. We know that all four cables were tensioned to 100 tonnes (98.421 tons), so the total tension (R) of a single high-level tie suspension bridge comprising two cables is 196.842 tons. The weight being supported by the main towers equals twice the vertical component (RV) of the tension in the cables, which equals 81.86 tons.

We know that, prior to 1960, each footway was carrying 78 tons of the weight of its high-level tie, so we can confidently say that the high-level tie suspension bridges are carrying the whole of the dead weight of the ties. The downside of this suspension bridge design, which had to deploy a very shallow angle of catenary, is that each main tower is subjected to a horizontal force of 385.1 tons acting at the level of the cable anchorages, *ie* *c*217ft above the base of the pier foundations. *(Author)*

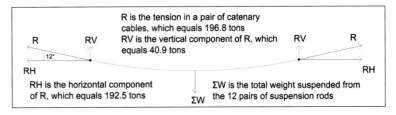

R is the tension in a pair of catenary cables, which equals 196.8 tons
RV is the vertical component of R, which equals 40.9 tons
RH is the horizontal component of R, which equals 192.5 tons
ΣW is the total weight suspended from the 12 pairs of suspension rods

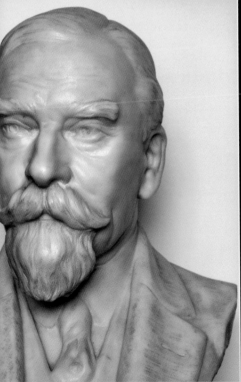

Chapter Eight

The men who built the bridge

The list of the men who designed and built Tower Bridge reads like a *Who's Who* of Victorian engineering, containing veritable titans of bridge design and construction, of contracting and of manufacturing industry. This chapter celebrates the achievements of these great men.

TOP LEFT Sir John Wolfe Barry *(ICE, photo by author)*
TOP RIGHT Baron Armstrong of Cragside *(National Trust)*
BOTTOM LEFT Sir John Jackson *(Edinburgh University Library)*
BOTTOM MIDDLE Sir William Arrol *(Historic Environment Scotland)*
BOTTOM RIGHT Sir Herbert Henry Bartlett of Perry & Co. *(UCL, photo by author)*

Sir John Wolfe Barry
KCB FRS

(7 December 1836–22 January 1918)

The fifth and youngest son of Sir Charles Barry RA, designer of the new Palace of Westminster, John Wolfe Barry was educated at Glenalmond School in Perthshire and at King's College, London. After a spell of practical training in the shops of Lucas Brothers, he became a pupil of John (later Sir John) Hawkshaw. Hawkshaw employed Barry as assistant resident engineer on the Charing Cross railway bridge project and as resident engineer on the Cannon Street railway bridge and station project.

In 1867, Barry started his own practice, taking on Henry Marc Brunel as partner 11 years later.

In April 1874, Barry married Rosalind Grace Rowsell of Hambledon. They had four sons and three daughters.

Barry and Brunel designed the wrought-iron St Paul's Railway Bridge (now known as Blackfriars Railway Bridge), built by Lucas & Aird and opened in 1886. Tower Bridge was their next major project.

Cuthbert A. Brereton joined the practice in 1892 and, with Brereton, Barry designed Kew Bridge, a granite arch bridge built by Easton Gibb & Son which opened in 1903.

Barry was also very active in the design of docks, including the Lady Windsor Deep Lock and graving dock at Barry, a new entrance to the Tyne Docks, Limekiln Wharf, the Alexandra Dock at Newport, the Immingham Dock and graving dock, the Royal Edward Dock at Avonmouth and extensions to Grangemouth Dock, Surrey Commercial Docks and the Middlesbrough Dock.

Sir John was also active on railway projects both at home and abroad.

In 1897 he became Knight Commander of the Bath, having been granted Companionship of the Bath in 1894 on the day Tower Bridge opened.

Throughout his life he was active in promoting technical education and the engineering profession, serving as President of the Institution of Civil Engineers from 1896 to 1897. He was a man of great charm, tact and selflessness, with a prodigious capacity for work. He served on numerous commissions and Royal Commissions, and acted as arbitrator on many contracts. He was intimately concerned with the establishment of the National Physical Laboratory.

At the turn of the 20th century, he urged the council of the Institution of Civil Engineers to form an Engineering Standards Committee, the objective being to standardise iron and rolled-steel sections. Two members each from the Institution of Civil Engineers, the Institution of Mechanical Engineers, the Institution of Naval Architects and the Iron and Steel Institute first met in 1901. The Institution of Electrical Engineers joined in 1902, the Kitemark being registered on 12 June 1903. The name was changed from the Engineering Standards Committee to the British Standards Institution in 1931.

He served as chairman of Cable and Wireless from 1900 to 1917.

Sir John was widely acknowledged as the head of his profession and his death was seen as a great loss to the nation and to the British Empire. He is buried at Brookwood Cemetery.

Sir John Jackson
CVO FRSE LLD

(4 February 1851–14 December 1919)

John Jackson was a celebrated civil engineering contractor, specialising in dock, harbour and waterway construction. He was born in York, the youngest of the five children of an elderly goldsmith, and educated at Holgate Seminary. At the age of 15, he was apprenticed to engineer William Boyd of Messrs Thompson & Boyd of Spring Gardens Engineering Works, Newcastle upon Tyne. In 1869, he went on to study engineering and philosophy at Edinburgh University under Professors Fleming-Jenkin and Tait.

Returning to Newcastle, he joined his brother William's building business. In 1875, at the age of 24, he started his own firm, John Jackson (Contractor). Within a year of starting his business he had delivered four successful contracts. His first large contract, won in 1876 and valued at £206,662 6s 0d, was for the building of Stobcross Docks in Glasgow (Contract No. 4). The fact that he had lost most of his hair by the age of 19 helped to sway the selection panel that he had the maturity to deliver this major project; he was only 25 years of age at the time. The contract was delivered nine months late, but he was still learning his trade; by the age of 30 he had made his name.

Jackson married his cousin Ellen Julia Myers of Lambeth on 31 May 1876. They had three sons and six daughters, but only five daughters survived. In February 1880, the Jacksons moved to London, taking a house at 26 Holland Villas Road, near Holland Park.

When he won Tower Bridge Contract No. 1, he was also completing Wicklow Harbour and engaged on a c£200,000 contract at Middlesbrough docks.

After winning Tower Bridge Contracts 1–3 and completing the work, Jackson went on to construct the last 8 miles of the Manchester Ship Canal in two-thirds of the permitted contract time, this work being completed in 1894 and earning him a knighthood in 1895. He made extensive use of steam shovels on this contract and achieved the highest productivity of any of the contractors employed on the project.

In 1894, the Jacksons moved to 10 Holland Park, a large property with 12 bed and dressing rooms, four reception rooms, extensive 'domestic offices', stables, a coach house and a tennis court. He had arrived.

Contract after contract was completed successfully. He proved himself to be an excellent estimator, a sound businessman and an inspirational leader, who was committed to the welfare of his workmen.

His largest project in the UK, valued in the region of £4 million, was the extension of the Admiralty Works at Keyham, Devonport, between 1896 and 1907. During this time, Jackson and his family leased Pounds House, a large house then on the outskirts of Plymouth. They also retained the London house.

He was elected a Fellow of the Royal Society of Edinburgh for services to engineering. His company was renamed and registered as Sir John Jackson Limited on 17 August 1898.

He stood unsuccessfully for the post of Member of Parliament for Devonport in both 1904 and 1906. He was finally elected in 1910, serving the constituency until 1918.

Other works in the UK included the extension of the Admiralty Pier at Dover, the Commercial Graving Dock and the Deep-Water Lock at Barry, the extension of the Prince of Wales Dock at Swansea, the reconstruction of the North Breakwater at Tynemouth and the breakwater and docks for Burntisland Harbour. On some of these contracts he renewed his working relationship with John Wolfe Barry.

He also undertook engineering work abroad, having at various times subsidiary companies in Bolivia, Canada, Chile, South Africa and Turkey. He owned a shipping line, Westminster Shipping Co. Limited, which he used to transport machinery and materials to overseas contract sites.

Overseas work included the building of a graving dock in Simonstown, South Africa,

BELOW Sir John Jackson. *(Edinburgh University Library)*

and harbour rebuilding works in Singapore. He built a breakwater in Victoria, British Columbia, and naval docks in Ferrol, Spain. He also built a railway from Arica in Chile to La Paz in Bolivia which crossed the Andes at 14,500ft. Another notable project was the building of a barrage across the Euphrates at Hindiyyah, during the course of which he diverted the river. He was even consulted as to the feasibility of constructing a bridge between Calais and Dover.

He became a wealthy man, owning a house in Belgrave Square and a country seat – Henley Park, near Henley-on-Thames.

From about 1910, he enjoyed the company of a mistress, Mrs Mabel Henderson of Hascombe Grange, near Godalming, Surrey. While at her house, on 14 December 1919, he had a fatal heart attack. His obituary avoided salacious detail and simply stated that he had died in Godalming. He left Mrs Henderson a large sum of money in his will and his wife a much larger sum.

BELOW No photograph could be found of William Webster, but his name appears on an iron casting in the cathedral-like interior of Crossness Pumping Station. *(Peter Dazeley)*

William Webster

(May 1819–1 February 1888)

William Webster was born in Wyberton, Lincolnshire, a small village about 2 miles outside the centre of Boston. He was educated locally and apprenticed to Mr Jackson, a Boston builder. The day after he was released from his apprenticeship, he started his own business. This tells us something about the man.

He worked on the restoration of St Botolph's parish church in Boston and, with Sir Gilbert Scott, undertook a complete restoration of the part 19th-century church of St Peter and St Paul, Algarkirk, some 6 miles south of Boston. This work was completed in 1851.

His work locally included the construction of the Corn Exchange (1855), the Reading Rooms and Britannia Oil Mill (1856) in Boston.

Between 1856 and 1860, he took on two huge contracts, the building of the County Pauper Lunatic Asylum for Cambridgeshire at Fulbourn and the Three Counties Asylum at Arlesey, Bedfordshire. Both of these building complexes are absolutely vast.

Moving to London in 1860, he undertook many major projects for Metropolitan Board of Works Chief Engineer, Joseph Bazalgette. These included work on the Southern Outfall at Crossness and the main interceptor sewers on both banks of the river. He built the Crossness Pumping Station (1865) and the Abbey Mills Pumping Station (1868).

In 1868, he was elected an associate member of the Institution of Civil Engineers in recognition of his outstanding contribution to the built environment; this was an honour for someone with no formal, tertiary-level engineering training.

For the MBW he built the whole of the Albert Embankment (1869) and Chelsea Embankment (1870) and a length of the Victoria Embankment (1870). He also undertook the extension of the embankment at the new Palace of Westminster and the building of the foundations for St Thomas's Hospital on reclaimed ground.

In 1873, he completed the construction of the gothic-style Dissenters' Chapel at Hither Green Cemetery. Other works included the building of a bridge in Maidstone, the construction of Holborn Viaduct station (1874) and the Station

Hotel (1877), Poplar Gasworks (1878) and the extensive waterworks at Hampton.

He built the Western Pumping Station in Pimlico (1875), which includes a 272ft-high chimney. The chimney is of Italianate design and comprises a tapered square brick outer tower encasing a circular smoke flue. Internal access is via a cantilevered spiral staircase naturally illuminated by vertical light openings. The chimney contains 25 flights of stairs and 13 landings.

Tower Bridge Contract No. 4 for the southern approaches, which included the accommodation needed to house the coal store, boiler rooms, engine rooms and accumulator tower required to generate the hydraulic pressure for the bridge, was let to his firm six months after his death. There can be no doubt that the firm at that stage owed everything to William Webster and that the contract was delivered just the way he would have wanted.

He died at the home which he had designed and built on Lee Terrace, Wyberton House. His eldest son, also William, ran the firm for a few years, undertaking the construction of the Blackheath Concert Halls in 1896.

Sir William Arrol

(13 February 1839–20 February 1913)

William Arrol was born in the village of Houston, near Paisley, Renfrewshire, the son of a spinner. At the age of nine, he started work at Coats' cotton mill, working in the turning shop and spool department. The most famous product of the mill was the Paisley-pattern shawl, which became fashionable after being worn by the young Queen Victoria. Arrol's parents moved to Paisley in 1850 where, at the age of 14, he was apprenticed to Mr Thomas Reid, a blacksmith. He studied mechanics and hydraulics at night school.

In 1863, he secured the job of foreman in the boiler works of Messrs Laidlaw & Sons in Bridgeton, Glasgow. He stayed for five years, learning his trade and saving up £85 to start his own boiler-making business. This he did in 1868, employing about 30 men. In 1872, needing more space, he built the first shops of his new Dalmarnock Iron Works, expanding the scope of his business to steel construction.

His first major project was for the ironwork

of a viaduct over the River Clyde at Bothwell, known as the Craighead Viaduct, for the North British Railway Company. Having delivered his tender for the ironwork to the prime contractor, Messrs Charles Brand and Son, he simply declined to leave the premises without the order. He devised a new method of undertaking work of this kind. The normal practice was to build bridge girders on trestles resting on piles driven into the river bed; his novel approach (of particular value for a bridge 120ft above the water) was to build the girders on the shore at one side of the bridge and then to roll them into place over the tops of the piers. The bridge was completed in 1875.

His next project was the building of the Caledonian Railway Bridge over the Clyde at Broomielaw. Work on the bridge, which was to carry the railway from the then terminus south of the river to Glasgow Central station, began in 1876. Because of the weight the bridge was designed to carry, the flanges of the wrought-iron girders were especially thick. His workmen sought additional remuneration for drilling and riveting these plates. Faced with this challenge, Arrol designed and built heavy-duty radial drills and hydraulic riveters, which allowed work to proceed rapidly and at reasonable cost. The hydraulic riveters operated at a pressure of 1,000psi and lengthy experiments were necessary to design and manufacture satisfactory flexible water supply hoses. The bridge opened on 1 August 1879.

In 1878, the Forth Bridge Company contracted with Arrol to build a bridge over the Firth of Forth to the design of Sir Thomas Bouch. The contract was cancelled after the collapse of Bouch's Tay Bridge.

In 1882 he won the contract to construct the new Tay Bridge to the design of William Henry Barlow. Construction started in mid-1883 and deployed 24,600 tons of iron and steel, 68,900 tons of concrete, 10 million bricks (weighing 36,900 tons) and 3 million rivets. No fewer than 14 men lost their lives during its construction, mostly by drowning. The bridge opened to traffic on 20 June 1887.

On 21 December 1882, the contract for the Forth Bridge was let to Sir Thomas Tancred, Mr T.H. Falkiner and Mr Joseph Philips, civil engineers, with William Arrol contracted to supply and erect the steelwork. Bridge engineers were Sir John Fowler and Sir Benjamin Baker. Preparing the machinery to tackle this daunting undertaking took a year and temporary plant cost £500,000. Once again, Arrol had to devise special drilling and riveting tools to facilitate the construction. An oil-fired rivet heater was also designed and 53,000 tonnes of steel were used.

The bridge opened on 4 March 1890 and William Arrol received a well-deserved knighthood on the same day. He was also awarded an honorary law degree by Glasgow University and the freedom of the Borough of Ayr. The bridge cost 73 lives, 38 through falling.

In May 1889, while still engaged on the Forth Bridge, he was awarded Tower Bridge Contract No. 6 to construct the superstructure of the bridge. The first payment on account for this contract was made on 2 July 1890.

Sir William went on to build a bridge over the Nile in Egypt, the Hawkesbury Bridge in Australia, the Keadby Bridge in Lincolnshire and the Warrington Transporter Bridge.

In 1895, Sir William was elected as Member of Parliament for the South Ayrshire constituency. He also served as President of the Institution of Engineers and Shipbuilders in Scotland during 1895 and 1896. He was elected a member of the Iron and Steel Institute in 1890.

His company was contracted by Harland and Wolff to construct a large gantry to enable the Belfast shipyard to construct the *Titanic*, the *Olympic* and the *Britannic* ocean liners. The image to the left shows the Arrol gantry surrounding two slipways. The gantry weighed 6,000 tons and measured 840ft long by 270ft wide by 228ft high. It also featured four electric lifts and a central revolving crane.

William Arrol married Elizabeth Pattison in July 1864, living in the Dennistoun area of Glasgow. Sadly, Elizabeth suffered from mental illness. In 1887, they moved to Ayr to escape the pollution of Glasgow. Arrol purchased a 50-acre estate and built Seafield House, in which he was able to indulge his love of literature, music and art. In 1889, a mental health nurse was employed to care for Elizabeth. She eventually died in 1904.

BELOW The Arrol gantry surrounding *Titanic* (*Olympic* is just out of frame on the right).
(Library of Congress)

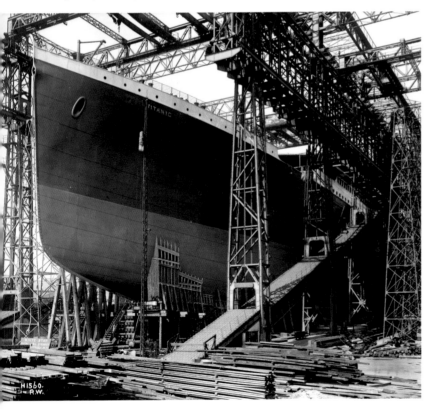

Fourteen months later, on 8 March 1905, Sir William married his cousin, Jessie Hodgart, who had been his companion on social engagements for some years. Jessie died just five years later and Sir William married for the third time on 16 October 1910. His new wife was Elsie Robertson from London, who was 35 years old. Today, one would be tempted to ask Elsie, 'What was it that first attracted you to the millionaire Sir William Arrol?' Sir William continued to take a close interest in his company and was often seen at Dalmarnock Iron Works, chatting to engineers and workmen about the challenges they were facing. He left an unsurpassed engineering legacy. His firm continued after his death, being acquired by Clarke Chapman in 1969.

Sir Herbert Henry Bartlett, Bart

(30 April 1842–28 June 1921)

Herbert Henry Bartlett was born in the village of Hardington Mandeville, about 4 miles south-west of Yeovil. He was the son of a carpenter. He attended local schools, moving to London at the age of 18 and taking rooms in Cheapside. He was apprenticed first to an architect and then to a civil engineer.

In 1865, Bartlett joined the building firm of John Perry, trading as Perry & Co. In 1872, John Perry took on his son William and Herbert Bartlett as partners. On 1 June 1876, Bartlett became managing partner. By February 1888, Herbert Bartlett was the sole proprietor of Perry & Co.

He married Ada Charlotte Barr on 18 April 1874. They had six boys and three girls between 1875 and 1892.

The company had offices at 56 Victoria Street and a 5-acre yard at Tredegar Works in Bow. The yard had a mechanised sawmill, timber-drying rooms, shops for joiners, smiths, fitters and plumbers, a paint shop and a stone yard equipped with a diamond circular saw and electric gantry cranes. Stabling for the firm's horses was also provided.

Perry & Co. took on all kinds of jobs including groundworks and foundations, private house construction, chambers (apartments), bridges, docks and wharves, piers and jetties, pumping stations, gas works, railways and railway buildings, warehouses, stabling, garages, schools, churches, hospitals, hotels and tunnels.

At the time Tower Bridge Contract No. 7 was let (May 1889), Bartlett was building a gas works at Ponders End, the clock tower at the People's Palace, a warehouse at Millwall, stabling and signal boxes at Waterloo station, engine sheds at Nine Elms, a relief sewer at Forest Gate, a house at 11 Portland Place, a road bridge in Surbiton, Kew Gardens Bridge and undertaking drainage work in Canning Town.

The company was incorporated in 1908, at which time Herbert was governing director, with his son Hardington Arthur Bartlett also listed as a director.

The 10 January 1913 issue of *The Building News* ran this piece on Sir Herbert, which neatly summarises the scope of his contribution to the built environment:

Sir Herbert H. Bartlett, as principal of the great firm of Perry & Co. (Bow) Ltd., has carried out a great many public works, such as hospitals, a few of which are here given; St Thomas's Hospital, the Magdalen Hospital, the Children's Hospital, the Naval Hospital, Chatham, and the London Hospital Extensions.

Among Government works, this firm has executed many for the Admiralty, the War Department, and the Office of Works, including barracks, civil buildings, etc.

Of public buildings, we can only here mention Burlington House, many for various learned societies, the People's Palace, and town halls, asylums, workhouses, banks, offices, flats, residences, etc., the briefest descriptions of which would occupy pages.

Railways (including the Bakerloo, with two tunnels under the Thames), dock works, railway stations, widenings, the Tower Bridge, and other bridges over the Thames.

Sir Herbert Bartlett has been President of the Builders' Institute, the London Master Builders' Association, and three times Master of the Worshipful Company of Patternmakers, many other duties having added to the record of a full and busy life.

Works not mentioned in the piece above include substantial works at Waterloo station, the extension of Somerset House, the Piccadilly Hotel, the Hotel Cecil, the Hotel Victoria and Her Majesty's Theatre, Haymarket.

The Piccadilly Hotel project is instructive. Foundations were dug to a depth of 40ft below street level to accommodate three extra floors, making a total of 12 floors. An artesian well was sunk to a depth of 400ft. The design, by Norman Shaw, features a decorative façade requiring 104,000ft^3 of Portland stone. It took Perry & Co. just 18 months to build and fit out the hotel.

In 1892, Bartlett moved from a house in Tredegar Square to Wimbledon, staying there only briefly, perhaps because of its distance from the works. In 1897 he completed a suburban mansion at 110 Clapham Common North Side, possibly intended for the family's occupation. In the event, he decided to rent the house to a director of the London & South West Railway, who remained a tenant until 1912. In 1900, Sir Henry took a large house at 54 Cornwall Gardens, Kensington.

He was a supporter of Sir Ernest Shackleton's first expedition to the South Pole. He was made 1st Baronet Bartlett of Hardington Mandeville on 7 February 1913.

In 1911, he gave £30,000 to the University of London to establish a school of architecture. This exists today as the UCL Bartlett Faculty of the Built Environment (The Bartlett). He undertook many works for the Southern Command of the British Army during the First World War, mainly in the vicinity of Winchester.

His son Herbert Evelyn Bartlett died on 8 August 1917. Tragedy struck again on 11 January 1920 when his son Hardington Arthur Bartlett, then managing director of Perry & Co., was swept off the deck of Belgian steamer *Pieter de Koenick* during a storm in the English Channel. Arthur was en route to Belgium to supervise the rebuilding of the town of Dinant after the First World War.

Sources agree that, while he was a kind and responsible employer, he was a private and somewhat reclusive man who was not fond of company and could even be described as socially awkward. He and his wife did not socialise and never went out if it could be avoided. He never owned a motor car. He usually dined alone, and on railway journeys he travelled first class, with his wife and family in a third-class compartment, further down the train. We do know that he was fond of the countryside and became a keen yachtsman, serving as Commodore of the Royal London Yacht Club from 1902 to his death.

Sir Herbert Bartlett died at his home in Cornwall Gardens and was buried at Highgate Cemetery, leaving an estate of £476,638. However, throughout his life he had bought corner sites in London and this extensive property portfolio was conveyed in a trust to his family outside of his estate.

Perry & Co. was run for a few years by his second son, Robert Dudley Bartlett, but he was not successful and the firm went bankrupt in 1926.

Baron Armstrong of Cragside

(26 November 1810–27 December 1900)

William Armstrong was the second child, and only son, of a corn merchant and local politician who had business premises on the Newcastle quayside. William was educated at a private school in Newcastle, then at Whickham and, from the age of 16, at the grammar school in Bishop Auckland. As a child, he had always been fascinated by machines, but his father insisted on him becoming a lawyer. He was duly taken on as a trainee by his uncle, who had chambers in London. Returning to Newcastle, he became junior partner in the firm of Messrs Donkin, Stable and Armstrong. In 1835, he married Margaret Ramshaw and built a new house at Jesmond Dene, close to the homes of his business partner, Armorer Donkin, and his parents. His marriage would last 58 years but be childless. He pursued his interest in all things mechanical in his spare time.

There is a (probably apocryphal) account that, while fishing, he observed a waterwheel and realised that a substantial amount of power was being lost. He wrote an article in the 29 December 1838 issue of *Mechanics Magazine* pointing out the great advantages which would accrue from an efficient method of employing the pressure of a column of water. At that time, he was considering utilising water main pressure in situations where the water supply was taken from high ground in the neighbourhood.

In the 18 April 1840 issue of *Mechanics Magazine*, he reported on the trials of a water pressure wheel, which he had designed, and which had been built at his friend Henry Watson's High Bridge works. The diameter of the wheel was only 2ft 3in, yet it produced 5hp from a water pressure of 57psi. He went on to propose that engines of this type could be used for all manner of purposes, including the working of cranes, lathes and printing presses and for haulage in mines.

In the article, Armstrong dismissed the idea of water pressure engines constructed along the lines of steam engines, with multiple cylinders and valve gear, yet he would later manufacture these in large quantities. In 1845, he proposed to the local council that he be permitted to build a crane on the quayside, powered by water pressure to load and unload cargo; this was largely funded by Armstrong and Donkin. The crane was a success and three more hydraulic cranes were installed, manufactured at Messrs Watson's works. In 1846, he became a fellow of the Royal

Society in recognition of his experiments in hydroelectricity.

In 1847, Armstrong founded the Elswick Engine Works, the firm producing cranes and warehouse lifts. In 1850, he perfected the hydraulic accumulator as a means of achieving a quantum leap in water pressure, the system allowing pressures of 750psi to be readily obtained using a water pump driven by a steam engine, without the need for extraordinarily high water-storage tanks. This invention allowed hydraulic cranes to be installed in any docks in the country, the cranes capable of lifting far heavier loads than his early prototypes. During this early period of his career he was granted several patents in the field of hydraulics, which was – along with steam – to dominate motive power in the second half of the 19th century. Many hydraulic power companies were established in ports and manufacturing cities to supply high-pressure water, the London Hydraulic Power Company commencing supply in 1884. His crane business thrived.

In 1854, having read of difficulties manoeuvring heavy field guns at the Battle of Inkerman in the Crimean War, Armstrong approached the Secretary of State for War to propose a lightweight, breech-loading 3-pounder gun with a rifled barrel. His proposal was well received and he was invited to submit guns for trials. In 1855, his first gun, a 3-pounder, gave excellent results, achieving the same range as the 68-pounder then in service with the British Army. In 1857, he submitted an 18-pounder for trial, and subsequently a 32-pounder, both of which were equally satisfactory. The Elswick Ordnance Company was set up and Armstrong was soon asked to produce larger field, siege and naval guns. In 1859 he gave his gun-related patents to the British Government and was knighted by Queen Victoria.

In 1863, Sir William purchased Cragside, an estate at Rothbury in Northumberland, and planned the landscaping of his new property. He began to step back from the day-to-day management of his business.

In 1864 the Elswick Ordnance Company and W.G. Armstrong & Co. were merged to form Sir W.G. Armstrong and Company.

In 1867, Armstrong's company began supplying guns for warships built by Charles Mitchell at the Low Walker shipyard. He also produced hydraulic engines for capstans.

In 1869 architect Richard Norman Shaw was commissioned to enlarge Cragside, which would be equipped with hydroelectric-powered electric lighting and hydraulic lifts.

In 1876, Armstrong's company completed the construction of a swingbridge linking Newcastle and Gateshead, the bridge powered by hydraulic engines. The first opening of the bridge after it became operational was for a ship en route to Elswick to collect a 100-ton gun for the Italian Navy. Four 100-ton naval guns supplied in 1878 were equipped with hydraulic engines to rotate the turrets. He supplied 150-ton dockyard cranes to dockyards throughout the world.

In 1882, Mitchell's shipyard and Sir William's company were merged to form Sir W.G. Armstrong, Mitchell and Co.

In 1887 Sir William was created Baron Armstrong of Cragside and took a keen interest in the Irish Question in the House of Lords. Tower Bridge Contract No. 5 was won in December 1887.

Sir W.G. Armstrong, Mitchell and Co. merged with Joseph Whitworth and Co. in 1897 to form Sir W.G. Armstrong Whitworth and Co.

His wife, Lady Margaret Armstrong, died on 2 September 1893. By then, Armstrong was one of the richest men in Europe, the Elswick works employing 11,000 people and covering 50 acres. He purchased Bamburgh Castle and restored it, while continuing to develop his Cragside estate. He received honorary degrees from Durham, Oxford and Cambridge and held office as the President of the Institution of Mechanical Engineers and the President of the Institution of Civil Engineers.

As a philanthropist, he supported the evolution of the University of Newcastle from the College of Physical Science, which he had founded. He built almshouses at Rothbury and donated the gorge of Jesmond Dene and nearby Armstrong Park to the people of Newcastle. He left £100,000 in his will to build a new Royal Victoria Infirmary in Newcastle.

After his death at Cragside, his company continued to expand, employing 30,000 people in 1914 and manufacturing airships, aeroplanes, guns, ammunition, cranes, swingbridges, capstans, docks, warships, submarines, motor vehicles and steam locomotives. The company supplied the guns for HMS *Dreadnought* (1906) and for 42 further Royal Navy battleships, the last being HMS *Vanguard*, laid down in 1942. Guns were also supplied to 19 battle-cruisers, including HMS *Hood*.

'However high we climb in the pursuit of knowledge, we shall still see heights above us, and the more we extend our view, the more conscious we shall be of the immensity which lies beyond.'
Lord William Armstrong

Armstrong bestrode 19th-century engineering like a colossus. His company continues to this day in the guise of BAE Systems.

Common threads

Reviewing the lifetime achievements of these men, one is struck by the sheer number of very substantial projects successfully carried out by each; projects which were undertaken far more quickly than they could be today, and which had a level of architectural and engineering complexity demanding great skill and the highest standards of workmanship.

They must have been excellent estimators and businessmen, man managers and leaders. More than this, they must have had the skill to appoint engineers, site managers and foremen capable of keeping very large workforces gainfully and efficiently employed.

These projects were costed without spreadsheets, pitched without PowerPoint® and managed without mobile phones or project management software. They were implemented predominantly by manpower and horsepower, with some steam and hydraulic equipment.

When we think of builders today, we might visualise someone dressed in a hi-vis jacket and hard hat, leaning against some Herras fencing and talking on a mobile phone – or possibly enjoying a cup of very strong tea. It wasn't like that in late Victorian times. The image I cannot escape is that of 5,000 ants building an anthill. Every ant is carrying something. Every ant is moving at the double. The worker ants wear cloth caps. The ants barking out the orders wear bowler hats. Every ant works ceaselessly until the job is completed. This is what the construction sites of these great contractors must have looked like.

With these men and their companies, Tower Bridge was in very safe hands.

Sir Horace Jones RIBA

(20 May 1819–21 May 1887)

It would be churlish to conclude this account of the men who built the bridge without a summary of the life and achievements of Sir Horace Jones. After all, the bridge was his child, even if he did not live to see it born.

Horace Jones was born at 15 Size Lane in the City of London, the son of a solicitor. At the age of 17 he was articled to architect John Wallen, qualifying as an associate member of the Royal Institute of British Architects at the age of 23.

He travelled to France, Italy and Greece to study classical architecture and established an architectural practice which he ran until 1864.

His first major commission was for Cardiff Town Hall, which was opened in 1854. He was then asked to design a grand house at Caversham Park, near Reading, for an iron magnate from Merthyr Tydfil. It is interesting that this design departed from normal practice by featuring a cast-iron frame.

In 1853, his Surrey Music Hall was opened in Walworth. The building, which again featured an iron frame, was designed as a performance space and conservatory, seating 10,000. Sadly it was to burn down in 1861. Other commissions included an office in Threadneedle Street, the Sovereign Life Assurance office in St James's Street, the Metropolitan (now Royal Free) Hospital and several department stores.

On 26 February 1864, at a meeting of the Court of Common Council, Horace Jones was elected to serve as Architect and Surveyor and Clerk of the City's Works, the voting being 95 votes in favour of Jones and 82 for the other candidate, Richard Bell.

One of his first responsibilities was the completion of the City Lunatic Asylum at Stone, near Dartford (later known as the Stone House Hospital). Horace Jones's other work for the City of London included Smithfield, Leadenhall and Billingsgate Markets, the restoration of the Guildhall, the Guildhall School of Music and several police stations. The design for Smithfield incorporated a double-glazed roof to reduce solar gain, plus a plethora of iron girders and wrought-iron stanchions.

On 15 April 1875 he married Ann Elizabeth Patch, with whom he had a son. He served as President of the Royal Institute of British Architects in 1882 and 1883. This was a matter of great personal satisfaction as some of his projects had not received universal acclaim, some feeling that his designs were rather too ornate and Italianate.

On Saturday 31 July 1886, Mr and Mrs Jones travelled to Osborne House on the Isle of Wight for Horace to be knighted by Queen Victoria for his distinguished services to the City of London as Architect to the Corporation.

Sir Horace died of heart disease at 30 Devonshire Place at the age of 68 years and one day. He left over £20,000, three houses and other properties to his wife and son.

Chapter Nine

Bridge maintenance

This manual has focussed primarily on the bridge as built. It's now time to move the story on and look at the upgrades which have been implemented to ensure that the bridge continues to ably meet the needs of road users, pedestrians and shipping. Will its new neighbour, the Thames Tideway Tunnel, have an impact on the bridge?

OPPOSITE In this view of the bridge taken in 1946 the high-level walkways are open to wind and weather. It tells a sad story of Britain and post-war austerity. Other aerial images of the bridge prove that this situation lasted for at least 6 years and possibly as many as 15. In 1960 the high-level tie suspension bridges were installed, the internal cross-bracing was removed and the walkways were re-roofed. *(Historic Environment Scotland EAW000650)*

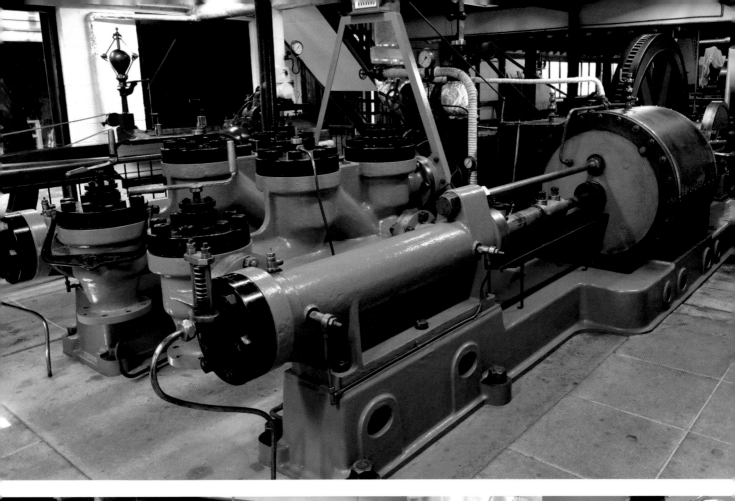

This chapter takes the form of a timeline, tracking the major maintenance, restoration and technology upgrades from 1894 to 2019.

The author is indebted to the authors (L.W. Groome, W.I. Halse, E.M. Longton and D.L. Stephens) of a paper on the bridge presented at a meeting of the Institution of Structural Engineers on 14 March 1985. This paper has provided much detail regarding the operation, maintenance, restoration and technological upgrades of the bridge from the time it opened on 30 June 1894 to the mid-1980s. A further paper, *Extending the Useful Life of the Tower Bridge in London*, by R.P. Stanley, C. Snowden and J.D. Hayward, believed to have been written in 1995, has also been useful in this regard.

Description of the main works since the opening of the bridge

1910
- The high-level walkways were closed to the public due to lack of use.

1928
- A major refit was undertaken to alleviate the effects of corrosion on the steel structure.

1941
- To provide increased availability, and because both main engines were in need of a major overhaul, a third (smaller) steam engine and hydraulic pump set was installed in the workshops to the east of the main steam engines and pumps. This space is now used as exhibition space, the footprint of the 1941 engine still being visible on the floor. This engine was built and installed by Vickers-Armstrong Limited. It had an indicated horsepower of 150hp at a steam pressure of 60psi, 42% of the power of each of the main steam engines.
- The engine was of the horizontal cross compound type, with a one-piece 9ft-diameter flywheel, a single 22in-diameter high-pressure cylinder on one side of the engine and a single 38in-diameter low-pressure cylinder on the other side. The stroke was 30in. Two hydraulic pumps of 6.625in-diameter were driven by the piston tail rods.

OPPOSITE **This is the 1941 Vickers-Armstrong engine at the Forncett Steam Museum. In the pump rear cylinder covers can be seen the engraved words 'Tested at 2,100psi'– an awesome pressure.** (*Both author*)

- The General Arrangement (GA) drawing shows that the rated speed was 60rpm. At this speed, the pumping rate would have been 71.82ft^3/min (447.3 gallons per minute). The GA specifies a slightly conservative pumping rate of 430 gallons per minute.
- After installation, the 1941 engine was used almost exclusively. At the governed speed of 28rpm, the time taken to replenish exhausted accumulators was 10 minutes 27 seconds.
- When the technology of the bridge was updated between 1974 and 1976, the 1941 engine and pump set was acquired by Dr Rowan Francis, who disassembled the engine, craned the component parts through the workshop roof, transported them to Norfolk and rebuilt the engine – an astonishing achievement.

Mid-1940s
- Although the bridge survived its first 50 years well, corrosion of the steel was a continual problem, particularly so on the high-level walkways, which were neither glazed nor watertight. Ornamental cast-iron panels on the walkways prevented access to the girder to which they were attached for inspection and painting. By 1942, all cast-iron panels had been removed.
- Mott, Hay & Anderson (MH&A) presented a report on the condition of the high-level footways to the Bridge House Estates Committee on 18 February 1943. It made for grim reading. The condition of the four footway girders was summarised as follows:

Girder A (upstream) Main tie in bottom boom	At many places severe corrosion of webs of channels of bottom boom.
	Some corrosion, not yet serious, of top flange plate of top boom; many rivet heads badly corroded in top boom.
Girder B	Severe corrosion, at a few points, of webs of channels of bottom boom.
Girder C	Severe corrosion at one point, less severe corrosion at a few other points, of webs of channels of bottom boom.
Girder D (downstream) Main tie in bottom boom	Some corrosion of webs of channels of bottom boom, but less severe than in other girders.

OPPOSITE

Replacement equipment in the east Surrey plant room: a Rexroth hydraulic power pack (top) and a MacTaggart Scott hydraulic motor/ gearbox set (bottom) driving the lower Surrey final driveshaft.

(Both author)

- MH&A calculated that the maximum stresses were in the cantilevered outer footway girders, the top booms of which were under a tensile stress of 6.15 tons/in^2 and the bottom booms of which were under a compressive stress of 4.9 tons/in^2 (both figures assuming no loss of section). MH&A noted that the usual working stress for mild steel was 7.5 tons/in^2 in tension and 5 tons/in^2 in compression. Assuming a 25% loss of section in the worst places, MH&A advised that the actual tensile stress in the top booms of the cantilevered outer girders would be 8.2 tons/in^2, while the maximum compressive stress in the lower booms of these girders would be 6.55 tons/in^2.

- The report notes the particular difficulty of repairs to the bottom booms of the outer girders due to the presence of the high-level ties. Water had collected inside the bottom booms and, though drain holes had been provided, 'the condition of affairs could not be looked upon without concern'. The key recommendations were that no repairs should be undertaken until after the war, and that further work was needed to specify the repair work and the means of its execution.

- A second MH&A report detailing the repairs needed was presented on 2 November 1943, together with two drawings. Two alternative schemes were presented. Scheme A involved adding steel plates to the bottom booms in places where the corrosion was severe, leaving the corroded members in place. Scheme B (preferred) involved the removal and replacement of corroded channels. Both schemes would require temporary steelwork to support the cantilevered footway girders, Scheme B requiring the steelwork to relieve the cantilevered footway girders of *all* their self- and imposed load.

- This report was discussed at the 10 December 1943 meeting of the Sub (Engineering) Committee of the Bridge House Estates Committee. The meeting would also consider alternative repair proposals put forward by Messrs Sir John Wolfe Barry and Partners.

- No information has been found which sheds light on which of the competing proposals was implemented. One would hope that repairs of some kind were undertaken before the situation as shown in the photograph at the top of page 144 was reached. Or was this the scheme proposed by Messrs Sir John Wolfe Barry and Partners?!

1960

- Corrosion of the high-level walkway girders continued to give concern. In order to relieve each outer girder of the c78-ton dead weight of the high-level tie which it was carrying, steel catenary cable 'suspension bridges' were installed, each featuring two cables – one inside and one outside of the walkway. The cables were anchored to the steel superstructure of the main towers. Vertical suspenders clamped to the cables were attached to the high-level ties. The previous suspenders connecting the high-level ties to the upper booms of the outer high-level walkway girders were removed. The lower booms of the outer girders were strengthened by the addition of a steel 'top hat' section which enclosed the high-level tie.

1962

- There had been continual problems with water ingress into the bascule roadways. This caused the swelling of the timber used

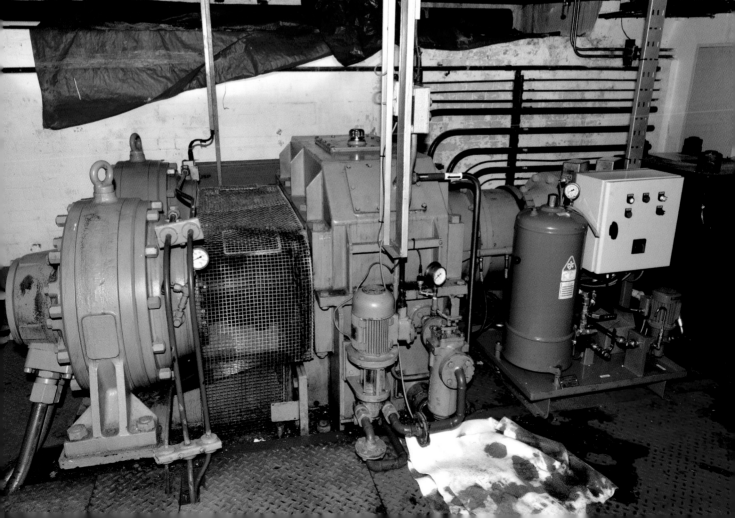

to infill the corrugated steel plates supporting the roadway and also caused the weight of the bascules to increase. Several of the corrugated steel plates were replaced and foamed polyurethane with a load-bearing performance of 1,000psi used to replace the timber infill. The work was done by C.J. Sims Ltd. The blocks, which were resistant to acids, benzene, paraffin, oil and petrol, were made by the Baxenden Chemical Co. Ltd. The road surface was of wood, with a top layer of calcined bauxite grit, embedded in epoxy resin.

Late 1960s
■ Vehicle-induced vibration in both main towers and the Surrey abutment tower was becoming a major problem due to heavier vehicles and increased traffic. These vibrations were tuned out by installing steel bracing between the roof framing and the attic floor of each main tower and by building block walls within the Surrey abutment tower. The Middlesex abutment tower was much less affected by vibration and no special measures were necessary.

1974–76
■ During this period a major technology upgrade took place. Consultants Mott, Hay & Anderson were commissioned to study options for replacement motive power for the bascules and to supervise its installation.

BELOW This is one of the 1976 operating desks. *(Fairfields Control Systems)*

Three different technologies were deemed suitable: an electrohydraulic system with electrically driven pumps and oil-hydraulic motors; thyristor-controlled electric motors; and a Ward Leonard system comprising an AC electric motor driving a variable-output DC generator, the output of which feeds a DC drive motor.

■ Tenders were invited, the system which was accepted featuring electrohydraulic technology. Each of the four plant rooms was equipped with:

1 A hydraulic power-pack powered by two 50hp AC electric motors, driving variable output Rexroth piston pumps, with a further (smaller) pump providing power to operate the pawls, the road gates and the nose bolts.
2 Two low-speed MacTaggart Scott hydraulic motors of 13,600 foot pounds of torque, continuous (18,000 foot pounds for 15 seconds). These are equipped with disc brakes and drive a common gearbox with a step-down ratio of 6:1, the output shaft of which is coupled to the bascule final driveshaft through a disconnecting coupling.
3 A motor starter and control/relay panel.

■ The equipment in one plant room was designed to be capable of fully lifting its bascule in 2 minutes. Normally, both sets of hydraulic motors on a pier are connected to their bascule final driveshaft.
■ A new 11kV electrical supply was taken from the south bank, routed along to the south tower, up and over a high-level walkway and down the north tower to the upstream accumulator void. Here transformers and switchgear were installed, providing a three-phase supply to the motor starter panel in each of the four plant rooms. A standby 415V three-phase supply is available at the north bank, together with a 135kVA standby generator which can open or close the bridge at half speed.
■ The upgrade was undertaken in two stages. In the first 12-month stage, the large hydraulic engines were removed from the two upstream plant rooms and the

new equipment installed. A three-month testing period followed, after which the two downstream plant rooms were upgraded. One of the small Armstrong hydraulic engines remains in each plant room; this was a requirement of the listed building consent for the technology upgrade works.

■ Operating desks were installed in the south-west watchman's cabin and the north-east operating cabin, retaining the old operating equipment in the south-east operating cabin.

■ A new electric lift was installed in each tower in the void previously occupied by a hydraulic lift, and the steel platforms and staircases in accumulator chambers were refurbished.

1977

■ The bridge, which had previously been painted chocolate brown, was redecorated in red, white and blue to celebrate Queen Elizabeth II's Silver Jubilee, all steelwork being repainted and all masonry cleaned.

■ The northern and southern approach roads were narrowed.

1978–79

■ Cast-iron balustrading on the shore spans, under the main chains, was removed, repaired and replaced, new sections being cast where repair was not possible. The edge trimmers of the shore spans, which were badly corroded, were removed and replaced with lightweight lattice girders with steel panels and GRP mouldings to faithfully reproduce the appearance of the original trimmers. The footways were reconstructed.

1974–82

■ Consultants were employed by the Corporation to determine how best to transform the bridge into a tourist attraction. A substantial amount of building work was undertaken to transform the engine and boiler rooms into a museum and bookshop. Improved lighting was installed throughout the main towers. Heating was installed in the main towers and high-level walkways. An additional fire escape staircase was constructed in each tower in a redundant lift shaft. Work on the walkways included the addition of external maintenance walkways,

GRP panelling to reproduce the original cast-iron dado panelling, new aluminium roofing, and new aluminium double glazing fitted to the openings in the lattice girders. GRP finials were installed on the corner pinnacles of the main towers, replacing the stone finials which had been removed for safety reasons during the Second World War.

1982

■ The bridge opened to the public for the first time since 1910, with a permanent exhibition inside called 'The Tower Bridge Experience'.

1984

■ Reconstruction and waterproofing of shore span carriageways.

1992

■ The bridge was redecorated in preparation for its centenary.

1993

■ The bridge was closed for 13 weeks while the roadway under the main towers was removed to enable the repair and redecoration of the underlying steelwork.

2000–1

■ A major project was undertaken to refurbish the hydraulic motors and replace the original, hard-wired, control system provided in 1976 to control the electrohydraulic technology. This control system was becoming unreliable. Some of the mechanical elements of the bridge, including the nose bolts, pawls and resting blocks, were also upgraded. The pawls were no longer correctly engaging and locking the bascules, and there was no system to ensure that the resting blocks on a pier shared the load of their bascule equally.

■ The project commenced with a feasibility study by systems integrator, Fairfields Control Systems. Fairfields, in conjunction with Bosch Rexroth, then proceeded to install a new electrical control system which incorporated a remote programmable logic controller (PLC) input/output subsystem in each of the four bridge machinery areas, with communication by means of fibre optic and coax cables to a central Rockwell

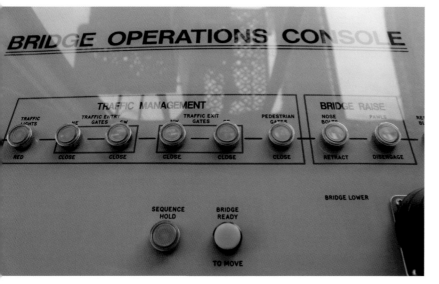

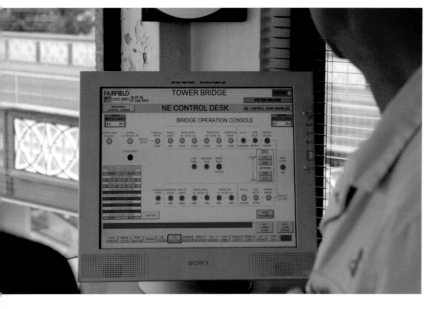

Automation ControlLogix PLC situated in the west Middlesex pier machinery room and linked to the two bridge control rooms and to the resting blocks. Each control room houses a supervisory control and data acquisition (SCADA) system that provides status alarms and condition monitoring. Refurbished control desks provide the primary operator interface. There are two rows of buttons, one for raising the bascules and one for lowering them. Each row of buttons is laid out in the sequence the operator/driver needs to follow to undertake bridge lifts and lowerings. The new system provides essentially the same interlocks as those provided by the original Saxby & Farmer interlocking lever frames. The images on this page show elements of the current control system. The new system is very much in line with the original duplication ethos of the bridge. Two desks are provided and there is a hot standby PLC, ready to take over control functions if the operational PLC fails.

■ The desk on the Surrey pier is normally used, the system ensuring that both bascules lift and fall precisely together – a joy to watch.

■ A major element of this upgrade was the introduction of active resting blocks to not only take an equal share of the weight of a bascule but also to relieve the main pivot bearings of much of the weight of the bascule, to reduce bearing wear. Each resting block is equipped with ABB Pressductor® load cells and duplicated analogue position transducers, which are driven by the PLC both to equalise weight distribution and for condition monitoring. The active resting blocks raise the bascule by approximately 10mm, taking all the weight off the main pivot bearings and suspending the bascule between pawls and resting blocks.

2004

■ A Vysionics speed and weight enforcement system was introduced to reduce the number of high-speed and overweight vehicles using the bridge. Vehicles in breach are photographed and their VRM is logged to enable the City of London Police to take enforcement action.

2008–11

■ The steelwork of the bridge was stripped of paint and redecorated in blue and white; 22,000 litres of paint were applied.

2012

■ In June 2012, the Olympic rings were suspended from the high-level walkways to mark the impending games.

2016

■ The bridge was closed to traffic for three months to allow maintenance to the bridge-lifting plant, to resurface the roadways and footways, to replace expansion joints and to waterproof the brick arches under the

approaches. The ABB Pressductor® load cell system was upgraded with digital electronics from ABB Measurement & Analytics.

■ The hydraulic oil heating subsystems were also upgraded. The existing immersion heaters were overheating the gearbox oil as there was no effective system in place to circulate the oil past the heaters. Dana Brevini engineers installed a replacement heater/filter system, supplied by C.C. Jensen. The lights controlling road and river traffic were also upgraded at this time.

ABOVE One of the hydraulic cylinders that drives an active resting block. *(Fairfields Control Systems)*

BELOW The Olympic rings. *(Getty Images 150238189)*

Stewardship of the bridge

We live in a time when funding for infrastructure maintenance is tight, both at home and abroad. The US federal, state and county authorities have spent far too little, for far too long, on their infrastructure, including bridges, of which there are some 614,000. The resultant damage caused to steel bridges through corrosion and wear is serious, around 9% of all bridges in the USA being structurally deficient. It is possible that some will fail, and it is likely that a larger number are beyond repair and will need to be replaced in the coming years. It is fortunate that Tower Bridge's maintenance is funded by Bridge House Estates, an organisation with substantial funds, so much so that it makes charitable donations of some £20 million each year.

The Department of the Built Environment within the Corporation of the City of London takes its responsibility for the ongoing maintenance of Tower Bridge very seriously, as can be seen from the comprehensive catalogue of work which has been undertaken since the bridge was opened.

The bridge is now a major tourist attraction and entertainment venue. The Tower Bridge organisation which runs it, while also employing the technicians required for the routine maintenance and operation of the bridge-opening machinery, also contributes immensely to the task of defining and implementing a list of necessary maintenance and improvement projects. It would seem that the bridge is in safe hands, to which the ongoing stewardship of the bridge can be entrusted.

Areas for study and possible attention

Monitoring and maintaining the condition of the steelwork

This issue will never go away. The lower booms of the inner footway girders are beginning to show evidence of corrosion, and the fixed girders and cantilevers inside the bascule chambers are in need of redecoration to prevent further corrosion.

Maintaining the balance of the bridge

In this operations manual we have looked at the most significant forces acting on the bridge. Some of these forces are 'in balance', causing no net moment about the base of the piers. These include the dead weight of the main tower superstructure of columns, girders and masonry cladding. The horizontal pull of the long chains is also balanced by the horizontal pull of the high-level ties, imposing no moment about the base of the piers, provided the rockers are free to rotate to equalise these forces.

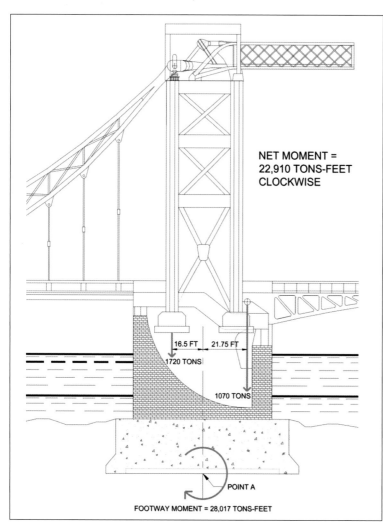

NET MOMENT =
22,910 TONS-FEET
CLOCKWISE

16.5 FT   21.75 FT

1720 TONS

1070 TONS

POINT A

FOOTWAY MOMENT = 28,017 TONS-FEET

LEFT The principal out-of-balance forces acting on the Middlesex pier in 1894 under no-load conditions. The net moment about Point A due to the moment of the footways and the moment of the two forces shown is a clockwise moment of 22,910 tons-feet. *(Author)*

'Out-of-balance' forces include the vertical applied load on the shore-side columns due to the suspension bridge, the weight of the bascule and the moment generated by the cantilevered footways. Second order out-of-balance forces include:

1　The weight of staircases, located next to the shore-side wall of each main tower.
2　The weight of additional brickwork on the shore-side of each pier.
3　The weight of the cantilevers supporting the fixed girders over the bascule chambers, acting near the river side of each pier.
4　The weight of the triangular girders carrying the main bearings, and the weight of the main bearings and shaft, acting near the river side of each pier.
5　The weight of the hydraulic engines acting towards the shore side of the pier.

We can be sure that Brunel and Barry took care to balance the bridge as best they could, as an out-of-balance bridge would cause uneven pressure on the London clay, which could cause pier rotation over a span of many years.

The Thames Tideway Tunnel

The Thames Tideway Tunnel is a massive project designed to provide a new, high-capacity deep sewer to supplement the capacity of Sir Joseph Bazalgette's low-level interceptor sewers – which are now running at full capacity – and thus prevent them from overflowing into the Thames. The project is being delivered by the organisation Tideway, which is owned by a consortium of investors.

Perhaps the biggest challenge faced by the bridge is that caused by the construction of the tunnel, the centre of which is some 36m below the river bed, the tunnel passing directly between the river piers.

RIGHT The principal out-of-balance forces acting on the Middlesex pier in 1894 under maximum design load conditions. Bascule live load to the left of the main pivot is ignored. The net moment about Point A is now a clockwise moment of 36,152 tons-feet. *(Author)*

As part of its duty of care to the City of London Corporation, studies were commissioned by Tideway to assess the impact of the construction of the tunnel on bridges.

Simulations and modelling undertaken by a multi-national engineering firm on the impact of the tunnel on Tower Bridge suggest that the construction of the tunnel will cause the piers to rotate very slightly towards each other. The effect at the level of the high-level ties will be that the Middlesex and Surrey towers will move some 24mm closer together (provided of course that the footway girders have the necessary clearance between them to permit this).

At first sight, this does not appear to be a very substantial issue. If the rockers on the bridge and abutment towers were working as designed, the suspension bridges could probably take this movement in their stride. As

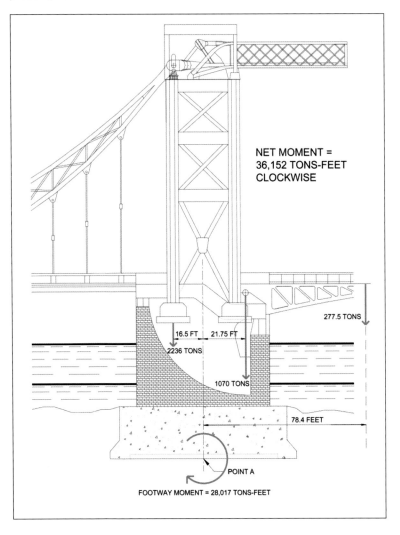

NET MOMENT = 36,152 TONS-FEET CLOCKWISE

277.5 TONS

16.5 FT　21.75 FT

2236 TONS

1070 TONS

78.4 FEET

POINT A

FOOTWAY MOMENT = 28,017 TONS-FEET

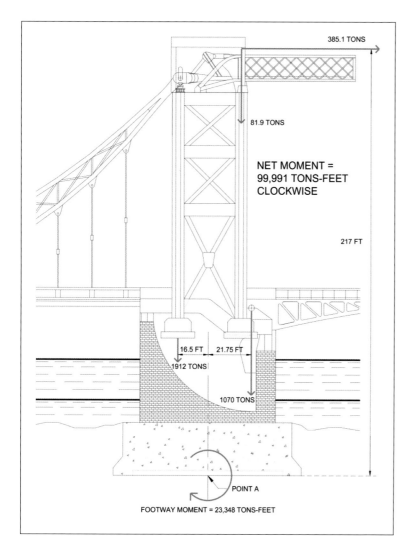

385.1 TONS

81.9 TONS

NET MOMENT = 99,991 TONS-FEET CLOCKWISE

217 FT

16.5 FT   21.75 FT

1912 TONS

1070 TONS

POINT A

FOOTWAY MOMENT = 23,348 TONS-FEET

the rockers are believed to have seized, the consultants concluded that each high-level tie will shorten by 24mm, reducing the tension in each tie from 8,939.3kN (897.2 tons) to 2,340.6kN (234.9 tons).

The effect of this reduction in tension in the high-level ties is that each main tower will be subject to an imposed horizontal force of 6,598.7kN (662.3 tons) applied by the large chains at their connections to the high-level ties (opposite top). This is a substantial force and the effect on the pier will be to cause an anti-clockwise moment (for the first time ever) of 33,794 tons-feet.

Unfortunately, the study failed to notice the presence of the two massive stiffening girders hidden inside the parapet of the Surrey shore span, which are fundamental to the way in which the suspension bridges operate. The stiffening girders are massive structural

members of the bridge, carrying between them a compressive load of some 1,000 tons. This error invalidates one of the conclusions of the study, namely that the shore spans would be able to 'hog' (lifting the roadway below the pins connecting long and short chains). The Middlesex shore span would be free to hog (if the pinned joints linking long and short chains were free to move) but the Surrey shore span certainly would not be, and the long chains would have to stretch, increasing the horizontal force acting on the main tower still further and also substantially increasing the vertical load carried by the shore-side columns. These forces would be additional to those shown in the image opposite top and would increase the anti-clockwise moment still further. These forces would cause unplanned stresses in several bridge components – the chains themselves, the main tower columns and the rocker assemblies. These latter are designed to carry enormous vertical load but are not engineered to carry large horizontal loads. Any bending of the columns would put unplanned forces on the masonry cladding of the bridge.

Tower Bridge is a world heritage site and, therefore, deserving of special protection.

The author has recently seen a Mott, Hay & Anderson drawing which shows the stresses in the main members of the bridge. It was certainly drawn prior to 1960 as no high-level tie suspension bridges are shown; it could even be from 1943, as stress numbers tally precisely

RIGHT An estimate of the out-of-balance forces acting on the bridge after the construction of the Tideway Tunnel, assuming that the rotation of the piers is as predicted by the consulting engineers.
(Author)

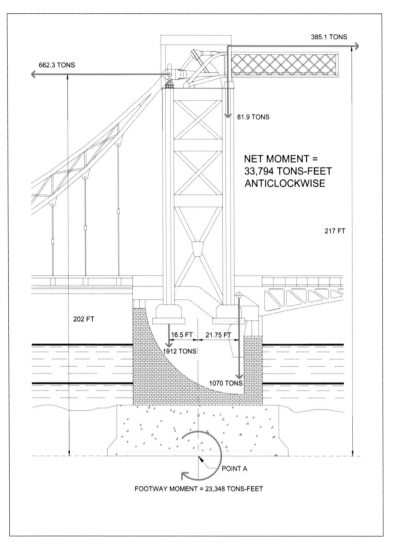

with those in the MH&A report of 18 February that year. A note in the drawing states that 'the bearings are rusted up'. So, it is highly likely that, 75 years later, they still are. There are two points of interest in the MH&A drawing:

1 The drawing notes that rusting of the bearings increases the stress at the bottom of the main columns to 8.15 tons/in^2, due to the pull of the long chains.
2 It would seem that MH&A also overlooked the stiffening girders!

The next 125 years

Should there be serious structural issues in the future, we can be grateful for the fact that the suspension bridges are at low level. This means that piers could be built at any time to provide support at the points where the long and short chains meet. This would be similar to the work done to provide support to Albert Bridge in the early 1970s and it would essentially preserve the look and feel of the bridge. Of course, it would result in the partial obstruction of two useful channels, much used by cruise and ferry boats.

An alternative approach – perhaps 100 years or more into the future when the shore spans need to be replaced – would be to rebuild the shore spans as concrete arch bridges, removing the chains and those heavy and problematic high-level ties; this would allow the bascule bridges to continue to operate. In another 100 years or so, perhaps the bascule bridges would need to be replaced by another concrete arch bridge, leaving just John Jackson's piers as a reminder of what once was. Let's hope that day never comes, and that Tower Bridge continues to lift our spirits and inspire us for many years to come.

RIGHT The main chains. *(Author)*

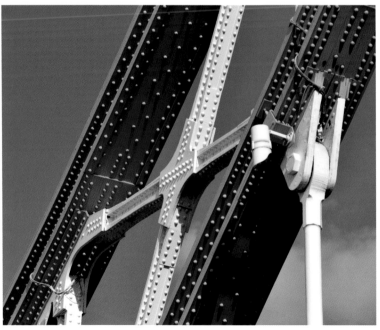

Appendix One

Suspension bridge force estimates

Objective

To model the force imposed on the column under each high-level tie via the rocker assembly, and the tension in each Middlesex and Surrey land tie and to compare these forces with the Cruttwell figures of 1,100 tons, 1,200 tons and 1,900 tons respectively.

Method

1 To set out the geometry of the two suspension bridges.
2 To calculate the weights of the long and short chains and the components making up the suspended shore spans. Imperial units are used throughout to mirror the calculations performed by Barry and Brunel. Dimensions and weights of the footways and roadway of the original 1894 bridge are used, but the current weight of each suspended shore span is also estimated.

3 To calculate the tension in the Middlesex abutment tower horizontal links and the high-level ties.
4 To calculate the load imposed on a main tower column by the rockers under a high-level tie.
5 To calculate the tension in the Middlesex and Surrey land ties.
6 To define what a fully loaded bridge might have been in 1894 and what a fully loaded bridge is in 2019.
7 To repeat the calculations for these cases. We will use a spreadsheet to do the calculations to four decimal points. Beware spurious accuracy!

Assumptions

For simplicity, it is assumed that:

1 The bridge is totally level.
2 End girders below the parapets have the same weight as longitudinal girders.
3 Each stiffening girder acts like a tie running in a direct line between the pin which connects the long and short chains, and the pin at the base of the abutment tower, so that no moment is applied to the pin connecting long and short chains.

BELOW The geometry of the Middlesex suspension bridge, showing the relative positions of the four pinned joints. The fact that the long and short chains are shown as straight lines does not invalidate the analysis which follows. *(Author)*

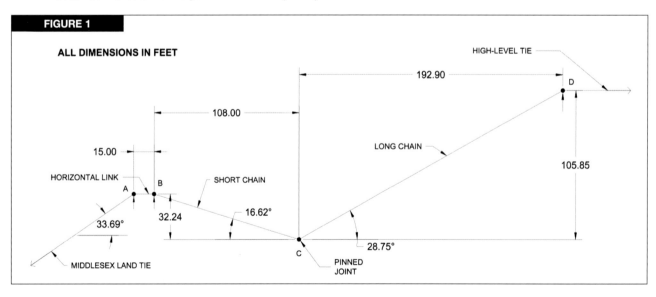

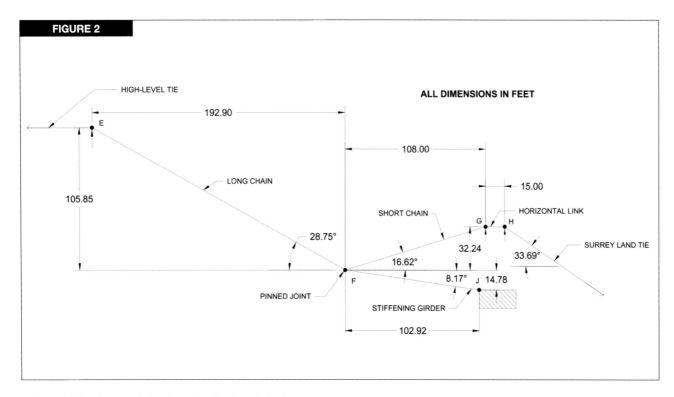

FIGURE 2

HIGH-LEVEL TIE

ALL DIMENSIONS IN FEET

192.90

E

LONG CHAIN

108.00

105.85

15.00

SHORT CHAIN

HORIZONTAL LINK

G   H

28.75°

32.24

SURREY LAND TIE

16.62°

33.69°

8.17° J 14.78

F

PINNED JOINT

STIFFENING GIRDER

102.92

4 Two-thirds of the weight of each stiffening girder is carried from the pin connecting the long and short chains, and one-third by the pin connecting the stiffening girder to the girder at the base of the columns in the abutment tower.

The geometry of the two suspension bridges

The geometry of the Middlesex suspension bridge is shown in Figure 1, and that of the Surrey bridge is shown in Figure 2. They differ only in that the Surrey bridge is fitted with stiffening girders, within both upstream and downstream

ABOVE The geometry of the Surrey suspension bridge, showing the relative positions of the five pinned joints and highlighting the asymmetry between the two bridges. *(Author)*

parapets. Note that, due to bending, we cannot assume that the tension in the chains will act in a straight line between the pins.

Calculated weights

From the drawing of the long chain available in Tuit's paper for *The Engineer*, it was possible to estimate the weight of the long chain, as shown in Table 1.

Component	Cross-sectional area (in²)	Length (ft)	Volume (in³)	Weight (tons)
TABLE 1: ESTIMATED WEIGHT OF LONG CHAIN				
Top boom (bays 6–10)	174.75	88.95	186,528.15	23.39
Top boom (bays 1–5)	208.20	112.57	281,650.14	35.29
Bottom boom (bays 6–10)	156.00	86.55	162,021.60	20.30
Bottom boom (bays 1–5)	199.50	123.15	294,821.10	36.94
Cross bracing	28.31	335.57	1144,009.91	14.29
Vertical bracing	28.31	84.86	28,831.19	3.61
Additional metal at nodes	**No. of nodes**	**Volume per node**	**Total (in³)**	**Weight (tons)**
Top nodes	9.00	11,114.38	100,029.38	12.53
Bottom nodes	9.00	14,633.63	131,702.63	16.50
Cross-brace nodes	8.00	2,798.50	22,388.00	2.81
Upper eye			150,860.00	18.90
Lower eye			95,698.00	11.99
Total				196.54

TABLE 2: ESTIMATED WEIGHT OF SHORT CHAIN

Component	Cross-sectional area (in²)	Length (ft)	Volume (in³)	Weight (tons)
Top boom				29.86
Bottom boom				29.14
Cross bracing	28.31	83.60	28,403.10	3.56
Vertical bracing	28.31	26.00	8,833.50	1.11
Additional metal at nodes	**No. of nodes**	**Volume per node**	**Total (in³)**	**Weight (tons)**
Top nodes	5.00	11,114.38	55,571.88	6.96
Bottom nodes	5.00	14,633.63	73,168.13	9.17
Cross-brace nodes	3.00	2,798.50	8,395.50	1.05
Upper eye				12.00
Lower eye				19.00
Total				111.85

No drawing was found for the short chain, so its weight was estimated using some of the parameters of the long chain, taking into account the reduced length of the booms and bracing, and the smaller number of nodes. This estimate is shown in Table 2.

The estimated total weight of a long and short chain pair is therefore 308.39 tons. The Schedule of Quantities and Prices in Contract No. 6, gives the total weight of steel in the booms, eyes, verticals, diagonals and nodes of *all* chains as 1,484 tons, or 371 tons per long chain/short chain pair. A figure of 370 tons is used in the analysis, with the ratio

of long-chain weight to short-chain weight taken from the total estimated weights in Tables 1 and 2. So the long chain weight is taken to be 236 tons, and the short chain weight is taken to be 134 tons. The weight of breeze concrete in the booms of a chain is of the order of 3–5 tons and has been ignored.

In the force analysis which follows, the suspension rods are numbered S1 through S14, as shown in Figure 3. The link between the pin connecting the short and long chains and carrying the transverse girder below is shown as SP.

The next task was to calculate, for the Middlesex bridge, the weight of the suspender rods, their screw couplings, the two thick plates between the lower two pins and the weight of the three pins in the assembly. The weight of the steel plate linking the pin connecting long and short chains to the transverse girder immediately below was also estimated. The figures are shown in Table 3.

BELOW The positions and numbering of the 14 suspension rods attached to the long and short chains. The Middlesex bridge is shown. The nomenclature is identical on the Surrey bridge. SP is the link suspended from the central pin linking the long and short chains. *(Author)*

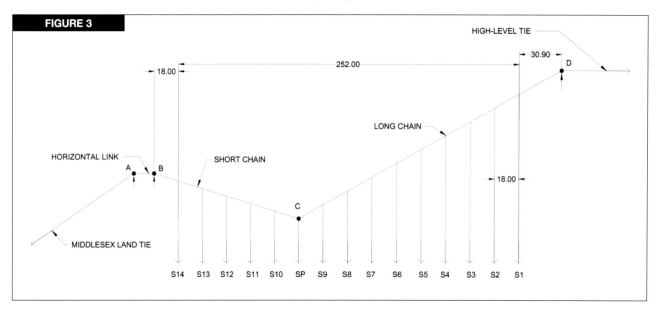

FIGURE 3

TABLE 3: WEIGHT OF MIDDLESEX SUSPENDERS, PINS AND PLATES, BY POSITION

Component	Cross-sectional area (in^2)	Length (ft)	Volume (in^3)	Weight (tons)	Weight of coupling, side plates and pins (tons)	Total
S1	28.27	76.20	25,850.09	3.24	0.71	3.95
S2	28.27	60.00	20,354.40	2.55	0.71	3.26
S3	28.27	46.20	15,672.89	1.96	0.71	2.67
S4	28.27	33.50	11,364.54	1.42	0.71	2.13
S5	28.27	25.40	8,616.70	1.08	0.71	1.79
S6	23.80	18.50	5,283.60	0.66	0.71	1.37
S7	23.80	13.80	3,941.28	0.49	0.71	1.20
S8	23.80	12.70	3,627.12	0.45	0.71	1.16
S9	23.80	12.70	3,627.12	0.45	0.71	1.16
S10	23.80	12.70	3,627.12	0.45	0.71	1.16
S11	23.80	12.70	3,627.12	0.45	0.71	1.16
S12	23.80	18.50	5,283.60	0.66	0.71	1.37
S13	28.27	24.20	8,209.61	1.03	0.71	1.74
S14	28.27	33.50	11,364.54	1.42	0.71	2.13
SP (Link under pin)						4.11

The 'as-built' weight of a complete suspended shore span was next estimated, using available information on footway and roadway width and the method of construction, using the figures below. Dividing the weight of a complete 270ft shore span by 15, gives the weight of an 18ft length of suspended span.

The weight of the shore span parapets is estimated from the Schedule of Quantities in Contract No. 6, which gives a total figure for cast-iron parapets for the shore spans and the high-level footways.

The weight of a transverse girder is given by Tuit as 'about 22 tons'. The Schedule of Quantities in Contract No. 6 suggests a figure of 28.75 tons. I have used a figure of 28 tons in the analysis.

TABLE 4: 1894 SHORE SPAN WEIGHT CALCULATIONS
Weight of one (270ft) suspended span in 1894

Component	Volume (ft^3)	Density (lb/ft^3)	Weight (tons)
Longitudinal girders*			177.50
Trough floor plates*			242.50
Breeze concrete fill of channels	3,788.26	82	138.68
Breeze concrete roadway	6,075.00	82	222.39
Wood block paving	4,860.00	45	97.63
Breeze concrete supports for paving	3,674.50	82	134.51
Services (estimate)			50.00
Paving	1,485.00	157	104.08
Parapets			200.00
Total			**1,367.30**
Weight per 18ft of suspended span			91.15
* Weights taken from Schedule of Quantities in Contract No. 6			

TABLE 5: 2019 SHORE SPAN WEIGHT CALCULATIONS			
Weight of one (270ft) suspended span in 2018			
Component	Volume (ft^3)	Density (lb/ft^3)	Weight (tons)
Longitudinal girders			177.50
Trough floor plates			242.50
Concrete fill of channels	3,788.26	150	253.68
Concrete roadway	5,314.90	150	355.91
Asphalt	1,063.00	150	71.18
Concrete supports for paving	2,852.59	150	191.02
Services (estimate)			50.00
Paving	1,906.00	150	127.63
Parapets			200.00
Total			**1,669.43**
Weight per 18ft of suspended span			111.30

Table 5 shows the estimated current weight of a suspended span. It relies on a recent study that sets out the dimensions and construction details of the current footways and roadway. If this recent study is correct, all the breeze concrete has been replaced with normal concrete.

It will be seen that the suspended weight of a shore span has increased by 302 tons (22%) since 1894, if the information in the study can be relied upon.

The weight acting on a column below the pin of a high-level tie due to the self-weight of the end of the tie, the rocker assembly and the pin and split sleeve is estimated, from the drawing, to be as follows:

Weight carried by rocker assembly under eye of high-level tie	12.00
Weight of rocker assembly under high-level tie	14.78
Weight of pin + sleeve at end of high-level tie	3.81
Total (tons)	30.59

The weight of a stiffening girder was estimated, from the drawing, to be 49.37 tons. The Schedule of Quantities for Contract No. 6, which *does include* the weight of stanchions, parapet girders and so on for stiffening the booms, puts this figure substantially higher, at 115 tons. I have used a figure of 100 tons in the analysis.

The original (1894) no-load weight suspended by each suspension point on the Middlesex and Surrey bridges is shown in Table 6. These figures include the weights of the suspension rods, screw connectors, side plates and pins. They also include half the weight of a transverse girder and half the weight of an 18ft section of shore span. The weight supported by the pin linking a long and short chain on the Surrey shore span assumes that two-thirds of the weight of the stiffening girder is supported by the pin connecting the chains and one-third by the abutment fixing.

We do not know the maximum design load used by

Barry and Brunel for each bridge. What we *do* know is that the load carrying capacity of each shore span was tested. These tests were conducted on one shore span at a time. Unlike the testing which took place on the Surrey lifting span, which used a real load imposed by traction engines with trailers loaded with granite, the testing was done by pulling down on the shore span using jacks on the staging below. Cruttwell states that a load of 1cwt per square foot of the shore span surface (roadway and footways) was applied. This is a huge load – 810 tons. Table 6 also shows the weight suspended by each suspension point on the Middlesex and Surrey bridges under this extreme load.

Finally, Table 6 also shows the weight suspended by each suspension point on the Middlesex and Surrey bridges under both 2019 no-load and assumed 2019 maximum-load conditions. The 2019 weight of each shore span has been estimated in Table 5. The maximum 2018 applied load has been estimated as the weight of the maximum number of 18-tonne, 14m-long vehicles that can be fitted on to a shore span, with a spacing of 2m between them. The maximum number of vehicles is 12, so the maximum imposed load is 216 tonnes, or 212.6 tons. Pedestrian load has been ignored.

I do not suggest that these figures can be totally relied upon as they are based on assumptions and weight estimates. However, they appear to show that the bridge today under no-load conditions is working some 16% harder than it did in 1894. The maximum possible load in 2019 is some 11% less than the maximum load in 1894.

Force analysis of the Middlesex suspension bridge

Figure 4 shows a simple force diagram for the east or west chains and ties of the Middlesex suspension bridge, as in

TABLE 6: FORCES ACTING ON THE LOWER BOOMS OF THE CHAINS

Suspension point	Weight supported by suspension points on Middlesex chains				Weight supported by suspension points on Surrey chains			
	1894 no load	1894 max load	2019 no load	2019 max load	1894 no load	1894 max load	2019 no load	2019 max load
S1	63.52	90.52	73.60	80.68	63.52	90.52	73.60	80.68
S2	62.83	89.83	72.91	80.00	62.83	89.83	72.91	80.00
S3	62.25	89.25	72.32	79.41	62.25	89.25	72.32	79.41
S4	61.71	88.71	71.78	78.87	61.71	88.71	71.78	78.87
S5	61.36	88.36	71.44	78.53	61.36	88.36	71.44	78.53
S6	60.95	87.95	71.02	78.11	60.95	87.95	71.02	78.11
S7	60.78	87.78	70.85	77.94	60.78	87.78	70.85	77.94
S8	60.74	87.74	70.81	77.90	60.74	87.74	70.81	77.90
S9	60.74	87.74	70.81	77.90	60.74	87.74	70.81	77.90
S10	60.74	87.74	70.81	77.90	60.74	87.74	70.81	77.90
S11	60.74	87.74	70.81	77.90	60.74	87.74	70.81	77.90
S12	60.95	87.95	71.02	78.11	60.95	87.95	71.02	78.11
S13	61.31	88.31	71.39	78.47	61.31	88.31	71.39	78.47
S14	61.71	88.71	71.78	78.87	61.71	88.71	71.78	78.87
SP	63.68	90.68	73.76	80.85	130.28	153.18	140.36	147.45
Total	924.01	1,329.01	1,075.13	1,181.43	990.61	1,391.50	1,141.73	1,248.03

1894, under no-load conditions. It assumes that points A, B, C and D are freely moving pinned joints preventing moments at these points, and that all rockers are unseized, lubricated and working as intended. For the bridge to be in equilibrium, the horizontal forces must add up to zero and so must the vertical ones. So, the tension in the Middlesex horizontal link (T_{MHL}) must equal that in the high-level tie (T_{HLT}):

Equation 1: $T_{MHL} = T_{HLT} = T$

The sum of the reactions at Points B and D must also add up to the total weight of the chains and their suspended loads, so:

Equation 2: $R_B + R_D = \Sigma$(suspended + chain weights) = $W = 1,294.01$ tons

BELOW Middlesex suspension bridge force diagram (1894, no imposed load). *(Author)*

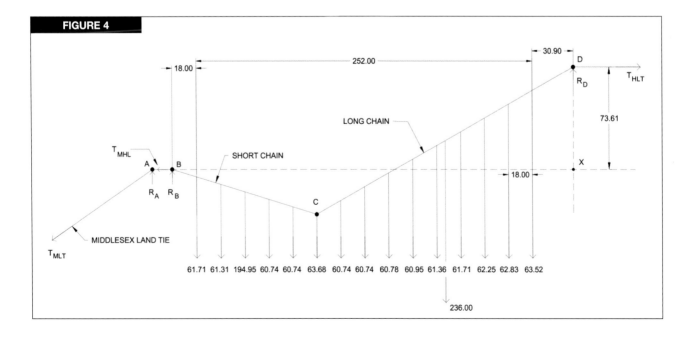

FIGURE 4

Taking moments about Point B (with suspended weights to four decimal places),

61.71*18 + 61.31*36 + 194.95*54 .... + 63.52*270 +73.61T
= 300.9R$_D$

Or 189,058.62 + 73.61T = 300.9R$_D$, from which:

Equation 3: T = 4.0878R$_D$ – 2,568.3904

Taking moments about Point C to the right,

60.74*18 + 60.74*36 + 60.78*54 .... + 105.85T = 192.9R$_D$

Or 73,081.59 + 105.85T = 192.9R$_D$, from which:

Equation 4: T = 1.8224R$_D$ – 690.4260

Eliminating T, using Equations 3 and 4,
4.0878R$_D$ – 2,568.3904 = 1.8224R$_D$ – 690.4260
From which R$_D$ = 828.99 tons
From Equation 3 or 4, T (= T$_{MHL}$ = T$_{HLT}$) = 820.31 tons
From Equation 2, R$_B$ = 465.02 tons
The force on the column under the high-level tie, F$_{COL}$ = R$_D$ + 30.59 = 859.58 tons

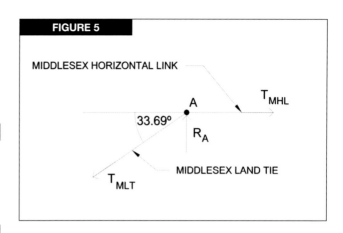

MIDDLESEX HORIZONTAL LINK

33.69°

A

T$_{MHL}$

R$_A$

MIDDLESEX LAND TIE

T$_{MLT}$

Figure 5 shows the forces acting at Point A

T$_{MLT}$*cos(33.69°) = T$_{MHL}$, or T$_{MLT}$ = T$_{MHL}$/ cos(33.69°) = 985.89 tons

R$_A$ = T$_{MLT}$*sin(33.69°) = 546.87 tons

We can now calculate the forces for the different load scenarios. These are shown in Table 7, taken from the spreadsheet.

TABLE 7: FORCES IN THE MIDDLESEX SUSPENSION BRIDGE				
Force	1894 no load	1894 max load	2019 no load	2019 max load
T$_{HLT}$	820.31	1,084.85	919.02	988.45
T$_{MHL}$	820.31	1,084.85	919.02	988.45
T$_{MLT}$	985.89	1,303.82	1,104.53	1,187.97
R$_A$	546.87	723.23	612.68	658.97
R$_B$	465.02	611.49	519.67	558.12
R$_D$	828.99	1,087.52	925.46	993.31
F$_{COL}$	859.58	1,118.11	956.05	1,023.90
W	1,294.01	1,699.01	1,445.13	1,551.43

We can see from the 1894 max-load figures that the 1cwt/ft^2 test load *was* the maximum design load, as our figure for the force acting on a column is very close to the Cruttwell figure of 1,100 tons. Our figure for the tension in the Middlesex land tie is some 100 tons higher than Cruttwell's figure of 1,200 tons. The key conclusion is that the bridge appears, in 2019, to be operating within its 1894 maximum design load envelope. The tension in the Middlesex land tie is a maximum of 1,188 tons, the maximum tension in the Middlesex horizontal link and the high-level tie is 988 tons, and the force on a main tower column under the rockers is 1,024 tons.

Force analysis of the Surrey suspension bridge

Figure 6 shows a simple force diagram for the east or west chains and ties of the Surrey suspension bridge, as built,

under no-load conditions. It assumes that points E, F, G and H are freely moving pinned joints preventing moments at these points, and that all rockers are unseized, lubricated and working as intended.

J is a pinned joint which can rotate, but no lateral or vertical movement is possible. In this section of the analysis, the 100-ton weight of the stiffening girder is assumed to be split equally between Point F and Point J.

Point F is effectively fixed in space by the stiffening girder, the short chain, the Surrey horizontal link and the Surrey land tie. The only movement possible at Point F is due to thermal expansion/contraction, or the lengthening or shortening of the components as the result of tensile or compressive forces.

Cruttwell informs us that T$_{SLT}$ is substantially larger than T$_{MLT}$, so we know that the stiffening girder is in compression. (The design of the stiffening girder enables it to withstand

FIGURE 6

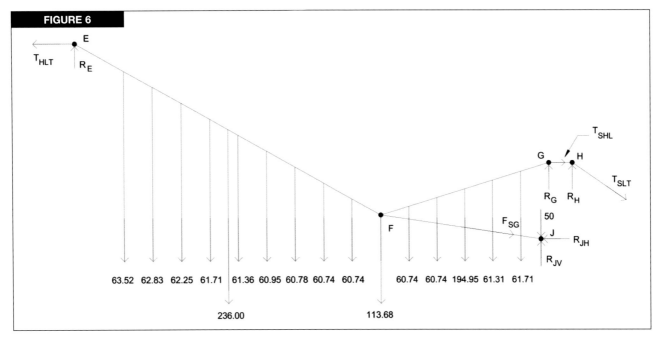

T_{HLT} E R_E

T_{SHL}

G H T_{SLT}

R_G R_H

50

F F_{SG} J R_{JH}

R_{JV}

63.52 62.83 62.25 61.71 61.36 60.95 60.78 60.74 60.74 60.74 60.74 194.95 61.31 61.71

236.00 113.68

compression, the suspension rods which pass through the girder preventing buckling.)

For the bridge to be in equilibrium, the horizontal forces must add up to zero, so:

Equation 5: $T_{HLT} + R_{JH} = T_{SHL}$

The vertical forces must also add up to zero, so:

Equation 6: $R_E + R_{JV} + R_G = \Sigma W + 50 = 1{,}396.95$ tons

(where ΣW is the weight suspended from the chains). R_E should equal R_D as both Middlesex and Surrey long chains share the same pull from the high-level tie. This was checked (by taking moments about Point F to the left) and found to be the case.

So,

Equation 7: $R_{JV} = \Sigma W + 50 - R_E - R_G = 567.96 - R_G$

Taking moments about Point J,
$188{,}886.55 + 5.08R_G + 120.63T_{HLT} = 295.82R_E + 47.02T_{SHL}$
Substituting known values for T_{HLT} and R_E yields:

Equation 8: $T_{SHL} = 906.20468 + 0.10803913R_G$

Eliminating T_{SHL} using Equations 5 and 8,
$T_{HLT} + R_{JH} = 906.20468 + 0.10803913R_G$
Substituting known value for T_{HLT},

Equation 9: $R_{JH} = 85.890961 + 0.10803913R_G$

FIGURE 7

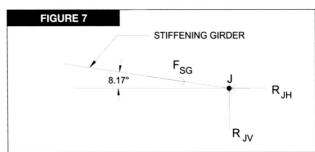

STIFFENING GIRDER

F_{SG}

8.17° J R_{JH}

R_{JV}

Figure 7 shows the forces acting around Point J.
$R_{JH} = F_{SG}*\cos(8.17°)$, so $F_{SG} = R_{JH}/\cos(8.17°)$
$R_{JV} = 50 + F_{SG}*\sin(8.17°) = 50 + R_{JH}*\tan(8.17°)$, so:

Equation 10: $R_{JV} = 50 + 0.14356777R_{JH}$

Eliminating R_{JV} using Equations 7 and 10,
$567.96 - R_G = 50 + 0.14356777R_{JH}$. So,

Equation 11: $R_{JH} = 3607.7869 - 6.9653516R_G$

Eliminating R_{JH} using Equations 9 and 11,
$85.890961 + 0.10803913R_G = 3607.7869 - 6.9653516R_G$
From which $R_G = 497.91$ tons.
From Equation 7, $R_{JV} = 70.05$ tons
From Equation 11, $R_{JH} = 139.68$ tons
$F_{SG} = R_{JH}/\cos(8.17°) = 141.12$ tons
From Equation 8, $T_{SHL} = 960.00$ tons
$T_{SLT} = T_{SHL}/\cos(33.69°) = 1{,}153.77$ tons
$R_H = T_{SLT}*\sin(33.69°) = 640.00$ tons
We can now calculate the forces for the different load scenarios. The spreadsheet results are shown in Table 8.

TABLE 8: FORCES IN THE SURREY SUSPENSION BRIDGE

Force	1894 no load	1894 max load	2019 no load	2019 max load
T_{HLT}	820.31	1,084.85	919.02	988.45
T_{SHL}	960.00	1,188.65	1,032.06	1,101.58
T_{SLT}	1,153.77	1,428.58	1,240.38	1,323.93
R_E	828.99	1,087.52	925.46	993.31
R_G	497.91	642.49	553.42	591.88
R_H	640.00	792.43	688.04	734.39
F_{SG}	141.12	104.87	114.19	114.29
R_{JV}	70.05	64.90	66.23	66.24
R_{JH}	139.68	103.80	113.03	734.39
W	1,346.95	1,744.91	1,495.11	1,601.44

T_{SLT} under 1894 maximum design load is 1,429 tons, a lot less than the Cruttwell figure of 1,900 tons. An explanation is needed!

I believe that the discrepancy can be explained by the geometry of the Surrey bridge. The structure to the right of Point F in Figure 6 is statically indeterminate. Any increase in T_{SHL} is balanced, ton for ton, by an equal and opposite increase in R_{JH}. Vertical reactions will adjust themselves to the increased tension in the ties and the Surrey short chains, and the increased compressive stress in the stiffening girders. The solutions in Table 8 are probably those appropriate to the situation in which there was no pre-tension in the Surrey land ties. The figure of 1,900 tons for T_{SLT} may have been the 1894 maximum load *design figure*, and the Surrey land ties could have been pre-tensioned to achieve this level of tension when the staging was removed from the completed bridge.

Plugging the value of 1,900 tons into the spreadsheet for 1894 maximum load:

$T_{SHL} = 1,900 * \cos(33.69°) = 1,580.90$ tons,
$R_H = 1,900 * \sin(33.69°) = 1,053.93$ tons,
$R_{JH} = T_{SHL} - T_{HLE} = 1,580.90 - 1,084.85 = 496.05$ tons,
$R_{JV} = 50 + R_{JH} * \tan(8.17°) = 121.22$ tons,
$F_{SG} = R_{JH}/\cos(8.17°) = 501.14$ tons
$R_E = R_D = 1,087.52$ tons,
$R_E + R_{JV} + R_G = \Sigma W + 50 = 1,794.91$ tons, from which:
$R_G = 1,794.91 - 1,087.52 - 121.22 = 586.17$ tons.

The Surrey land ties could have been easily pre-tensioned by installing the stiffening girders and the Surrey short chains (and at least the eyes of the long chains and the pins) and allowing the weight of these components to pre-tension the Surrey land ties.

Health warning

This simple analysis (which has been great fun!) is no substitute for a rigorous finite element analysis of the bridge, which would determine both forces and stresses in each member of the structure. Such a model would be extremely useful to the bridge engineer as a means of testing the impact of all future maintenance proposals. It would also be useful for exploring the potential impact of the construction of the Thames Tideway Tunnel on the structure of the bridge.

However, the analysis appears to indicate that the bridge in 2019, even under maximum loading, is still operating within its original maximum design load envelope.

The best thing is that, when we see a structural member such as the west Middlesex land tie shown on page 89, we now have some clue as to what is going on inside it; namely, that it is under a tension of between 1,105 tons and 1,188 tons, depending on the load on the bridge.

We also know that the Surrey short chains are under substantially more tensile stress than the Middlesex ones and are thus deserving of particular attention when assessing the effect of the Thames Tideway Tunnel on the structure of the bridge.

Appendix Two

Availability estimate for the bridge as built

We must begin with a 'health warning'. This analysis is not rigorous, as it is not based on real data on the various types of equipment faults encountered by bridge engineers and technicians between 1894 and 1977, and the average time taken to fix those faults. However, it should give us a feel for the level of availability delivered by the bridge with its original 19th-century hydraulic technology.

The availability (A) of an asset, be it a mainframe computer, a humble washing machine or a hydraulic engine, is normally calculated as:

A = MTBF/(MTBF + MTTR), where MTBF is the mean time between failures of the asset and MTTR is the mean time to repair the asset and return it to service when it does fail. Availabilities are essentially probabilities and can be manipulated as such. So, the probability of the asset being available (or 'up') is A, and the probability of it being unavailable (or 'down') is (1 − A).

A pair of assets can only be in one of four states. They are either both working, or Asset 1 is working and Asset 2 is down, or Asset 2 is working and Asset 1 is down, or both assets are down. The probability that they are both working is A_1*A_2, while the probability that both are down is $(1 − A_1)*(1 − A_2)$. The probability that Asset 1 is working and Asset 2 is down is $A_1*(1 − A_2)$, while the probability that Asset 2 is working and Asset 1 is down is $A_2*(1 − A_1)$. These probabilities must add up to 1 as there are no other possible states. So,

$$A_1A_2 + A_1(1 − A_2) + A_2(1 − A_1) + (1 − A_1)(1 − A_2) = 1$$

If the assets are identical, $A_1 = A_2$, so the equation simplifies to:

$$A^2 + 2A(1 − A) + (1 − A)^2 = 1$$

The probability that at least one of the two assets is available is $A^2 + 2A(1 − A)$. This is the same probability as one minus the probability that both assets are down, which is $1 − (1 − A)^2$, which is marginally easier to calculate. This is particularly so in the case of the hydraulic engines. The probability of all four engines in a pier being out of action would be $(1-A_{ENGINE})^4$, so the probability of at least one being operational would be $1 − (1 − A_{ENGINE})^4$, which is a lot simpler to calculate than all the other possible combinations.

It is a little more complicated when it is necessary to calculate the availability of (say) at least 3 out of 7 assets, but the binomial expansion comes to our aid. However, we will steer clear of that complexity in this simple model.

For Tower Bridge to be capable of operating with its original technology, it would have been necessary to have at least two out of the four boilers providing steam at working pressure, at least one steam-driven pump operational, at least one supply of high-pressure water to each pier, at least one out of the four hydraulic engines operational in the Middlesex pier, and at least one out of the four hydraulic engines operational in the Surrey pier. (A working driver's cabin on each pier and trained drivers would also be needed, but we will assume that these are available.)

So the availability of the bridge (A_{BRIDGE}) would be given by:

$$A_{BRIDGE} = (1 − (1-A_{BOILERPAIR})^2)*(1 − (1-A_{PUMP})^2)*(1 − (1 − A_{PIPE})^2)*(1 − (1 − A_{ENGINE})^4)^2$$

Note that, to avoid the binomial theorem, I have used the availability of a pair of boilers rather than any two out of the four boilers – a bit of a cheat.

It's time to estimate the ballpark availability of the several components. Let's start with the hydraulic engines. It is probable that the most frequent failures would have been due to minor problems such as leaking glands and seals, and blocked lubricators. These might have occurred every six months and taken four hours to fix. Problems like a seized valve or a bearing needing replacing may have occurred every five years and taken two days to fix. A really serious problem like a broken part in the valve gear might occur every 20 years and take two weeks to fix. (You see how arbitrary this analysis is!) Using these numbers, the Mean Time Between Failures would be 3,893.33 hours and the Mean Time to Repair would be 7.47 hours. This gives:

$$A_{ENGINE} = 3893.33/(3893.33 + 7.47) = 0.998085.$$

Lancashire boilers are unsophisticated, low-pressure boilers. They don't require a large number of boiler stays, which can leak. They don't have any boiler tubes (apart from the two large tubes containing the fires). Boiler tubes need replacing every few years and can also leak. The most likely failure would be the failure of a boiler fitting such as a pressure gauge, a sight glass or a valve, which might occur every two years and take four hours to fix, once the fire had been raked out and the boiler had been relieved of pressure, allowed to cool and drained. In practice, if a failure like this occurred, fires in a standby boiler would have been lit, probably taking between six and eight hours to raise steam. So the MTBF of a boiler would be two years, or 17,520 hours and the MTTR would be eight hours, making $A_{BOILER} = 0.999544$. Let's assume that the availability of a pair of boilers, $A_{BOILERPAIR}$, would be A_{BOILER}^2, or 0.999087.

A steam-driven pumping engine might have needed a gland repacking, a lubricator unblocking or a steam fitting replacing every year, all these repairs taking four hours. The pumps, operating as they do at extremely high pressure, might have needed attention every six months and would have been down for eight hours. A major problem might have occurred every ten years, requiring a week to fix. Using these numbers, the MTBF would be 2,825.81 hours and the MTTR would be 11.87 hours. This would make $A_{PUMP} = 0.995817$.

The availability of high-pressure water distribution to both piers is very hard to estimate. There were many individual lengths of pipe, each with a bolting flange at both ends. The individual lengths of pipe running up and down the steel columns in the main towers must have been short, as it would be impossible to get a long length of pipe into the column. The high-pressure supply pipes ran along the east side of the Surrey shore span, under the pavement. To accommodate movement under the pin linking the long and short chains, a joint in the pipes permitting a degree of movement was provided. Expansion joints were fitted on top of the east high-level footway. A large number of manual and hydraulically operated valves were included in the distribution system. The pressure was extraordinarily high, making leaks a high risk. I have a feeling that this might have been the Achilles heel of the bridge. Let's assume that there was a fault every six months and that it took 24 hours to repair. This would make the MTBF equal to 4,380 hours and the MTTR equal to 24 hours, from which an estimate of A_{PIPE} would be 0.994550.

Plugging these availabilities into the formula above, using a spreadsheet, yields two interesting results. Firstly, the spreadsheet calculates the availability of at least one operational hydraulic engine in both piers as 0.9999999999731, which is essentially 100% availability. Secondly, A_{BRIDGE} computes as 0.999951972 or 99.9952%. That might fall short of the level of availability of an air traffic control system, but it is an impressive achievement. Sir William Armstrong's design team is to be congratulated!

We have not taken into consideration the fact that the bridge had access, when needed, to high-pressure water from the London Hydraulic Power Company's mains. This would make the availability of the bridge even higher.

Appendix Three

Unit conversions

Length

1in = 25.4mm	1mm = 0.039370in
1ft = 0.3048m	1m = 3.280840ft
1yd = 0.9144m	1m = 1.093613yd

Area

$1in^2 = 645.16mm^2 = 6.4516cm^2$	$1cm^2 = 0.155000in^2$
$1ft^2 = 0.092903m^2$	$1m^2 = 10.763910ft^2$
$1yd^2 = 0.83612736m^2$	$1m^2 = 1.195990yd^2$

Volume

$1in^3 = 16.387064cm^3$ (/cc)	1 cubic centimetre = $0.0610237in^3$
$1ft^3 = 0.028317m^3$	$1m^3 = 35.314667ft^3$
$1yd^3 = 0.764555m^3$	$1m^3 = 1.307951yd^3$

Mass

1 Imperial (long) ton = 20 hundredweight (cwt) = 2,240lb	1 metric ton (tonne) = 1,000kg = 2,204.623lb
1lb = 0.453592kg	1kg = 2.204623lb
1 ton = 1.016047 tonnes	1 tonne = 0.984207 tons

Force

1 pound force = 4.4482216 newtons	1 newton = 0.2248089 pounds force
1 pound force = 1 pound (on Earth)	1 newton = 0.1019716kg (on Earth)
	1 kilogram = 9.806650 newtons (on Earth)

Pressure

	1 pascal = 1 newton per square metre
1psi (relative to atmosphere) = 6,894.7573 pascal	1 pascal (relative to atmosphere) = 0.0001450379psi
1 ton per square foot = 107,251.78 pascal	1 pascal = 9.323855×10^{-6} tons per square foot

Moment/torque

1 pound force-feet = 1.35581795 newton-metres	1 newton-metre = 0.73756215 pounds force-feet
1 tons force-feet = 3,037.032 newton-metres	1 newton-metre = 3.292688×10^{-4} tons force-feet

Principal sources and further reading

Welch, Charles, FSA (Librarian to the Corporation of London), *History of the Tower Bridge and of other bridges over the Thames built by the Corporation of London* (Smith, Elder & Co., 1894)

Tuit, James Edward, AMICE (Engineer to Sir William Arrol and Co), *The Tower Bridge – its history and construction from the date of the earliest project to the present time* (1894). The bulk of this book first appeared as an extensive article in the 15 December 1893 issue of *The Engineer* magazine.

'The History of Tower Bridge', a paper by David Lawrence Clackson of the Guildhall Historical Association, 1980.

'The Foundations of the River-Piers of the Tower Bridge', George Edward Wilson Cruttwell, MICE (resident engineer for John Wolfe Barry during construction), Paper No 2652, Proceedings of the Institution of Civil Engineers, 28 March 1893.

'The Tower Bridge: Superstructure', George Edward Wilson Cruttwell, MICE, Paper No 2938, Proceedings of the Institution of Civil Engineers, 10 November 1896.

'The Machinery of the Tower Bridge', Samuel George Homfray, MICE (the Sir W.G. Armstrong & Co design authority for the machinery), Paper No 2992, Proceedings of the Institution of Civil Engineers, 10 November 1896–97.

'The Tower Bridge', a lecture by John Wolfe Barry, MICE, published by Boot, Son and Carpenter, 1894.

Chamberlain's Cash Account (Tower Bridge Construction Ledger), 1886–93, London Metropolitan Archives, Ref: CLA/020/01/004.

Ledger, audited, purchase of property, rents received, erection of bridge, conservancy expenses, hire of tug etc (Tower Bridge Construction Ledger), 1887–1902, London Metropolitan Archives, Ref: CLA/020/01/005.

The Minute Book of the Corporation of London's Special Bridge or Subway Committee, London Metropolitan Archives, Ref: COL/CC/SBC/01/001.

Press advertisements for Tower Bridge contract tenders, London Metropolitan Archives, Ref: COL/CC/BHC/09/001.

Mott, Hay & Anderson reports of 18 February 1943 and 2 November 1943 into the condition of, and potential repairs required to, the high-level walkways of the bridge, including a stress analysis, London Metropolitan Archives, Ref: COL/PL/02/N/005/a, b, c and COL/PL/04/03/005.

London Metropolitan Archives, used generally as a source of contemporary alternative bridge proposals, Tower Bridge contracts, tenders, drawings, progress plans, hydraulic machinery invoices, miscellaneous documentation etc.

'Proposed Bridge over the Thames below London Bridge', a report by Sir J.W. Bazalgette CB to the Metropolitan Board of Works, 22 March 1878, The Science Museum Archives.

The John Wolfe Barry Archive (a collection of engineering plans for Tower Bridge from Wolfe Barry's office), The Science Museum Archives.

The Hardington Arthur Bartlett file, King's College London Archives.

The London Gazette, multiple issues.

The Engineer magazine, multiple issues.

Engineering magazine, multiple issues.

Graces Guide, multiple documents.

Proceedings of the Institution of Civil Engineers, multiple issues.

'Tower Bridge', a paper by L.W. Groome et al, *The Structural Engineer,* Vol 63A No 2, February 1985.

'Extending the Useful Life of the Tower Bridge in London', R.P. Stanley, C. Snowden and J.D. Hayward, International Association for Bridge and Structural Engineering, Symposium, Vol 73/1, pp83–88, 1995.

Matthews, Peter, *London's Bridges* (Shire Publications, 2008)

Assessment of the effects of tunnel-induced settlement on Tower Bridge, September 2013, Document Ref: 9.15.51, Report by Aecom for Thames Tideway Tunnel.

Index